Seismic Loads

Other Titles of Interest

Minimum Design Loads for Buildings and Other Structures, ASCE/SEI 7-10 (ASCE, 2013): Provides requirements for general structural design and includes means for determining dead, live, soil, flood, snow, rain, atmospheric ice, earthquake, and wind loads, as well as their combinations, which are suitable for inclusion in building codes and other documents. Includes a detailed commentary with explanatory and supplementary information.

Books Related to ASCE 7-10

Significant Changes to the Seismic Load Provisions of ASCE 7-10: An Illustrated Guide by S.K. Ghosh, Ph.D.; Susan Dowty, P.E.; and Prabuddha Dasgupta, Ph.D., P.E. (ASCE, 2010): Summarizes changes to the seismic provisions of ASCE 7-10 that might affect actual practice or enforcement, including the precise wording of the change.

Significant Changes to the Wind Load Provisions of ASCE 7-10: An Illustrated Guide by T. Eric Stafford, P.E. (ASCE, 2010): Translates changes to the wind load provisions of ASCE 7-10 into a form readily accessible by structural engineers, architects, contractors, building officials and inspectors, and allied professionals.

Snow Loads: Guide to the Snow Load Provisions of ASCE 7-10 by Michael O'Rourke, Ph.D., P.E. (ASCE, 2010): Illustrates key concepts and guides for applying the provisions of ASCE 7-10 to the design of new and existing structures that could collect falling or drifting snow.

Wind Loads: Guide to the Wind Load Provisions of ASCE 7-10 by Kishor C. Mehta, Ph.D., P.E., and William Coulbourne, P.E. (ASCE, 2013): Explains the wind load provisions of ASCE/SEI 7-10 as they affect the planning, design, and construction of buildings for residential and commercial purposes.

Books on Seismic Engineering

Earthquake Protection of Building Equipment and Systems: Bridging the Implementation Gap by Jeffrey A. Gatscher, Gary L. McGavin, and Philip J. Caldwell (ASCE, 2012): Offers a framework for applying the latest earthquake engineering research to the nonstructural elements of individual building projects, concentrating on mechanical and electrical systems.

Earthquakes and Engineers: An International History by Robert K. Reitherman (ASCE, 2012): Traces the evolution of humankind's understanding of the causes and characteristics of earthquakes and the development of methods to design structures that resist seismic shocks.

Guidelines for Seismic Evaluation and Design of Petrochemical Facilities by the Task Committee on Seismic Evaluation and Design of Petrochemical Facilities (ASCE, 2011): Presents practical recommendations regarding the design and safety of petrochemical facilities during and after an earthquake, including guidance on design details and considerations that are not included in building codes.

Seismic Evaluation and Retrofit of Existing Buildings, ASCE/SEI 41-13 (ASCE, 2014): Describes deficiency-based and systematic procedures that use performance-based principles to evaluate and retrofit existing buildings to withstand the effects of earthquakes.

Seismic Loads

Guide to the Seismic Load Provisions of ASCE 7-10

Finley A. Charney, Ph.D., P.E., F.ASCE, F.SEI

 **ASCE** PRESS

Library of Congress Cataloging-in-Publication Data

Charney, Finley Allan.
    [Seismic loads (2014)]
    Seismic loads : guide to the seismic load provisions of ASCE 7-10 / Finley A.
Charney, Ph.D., P.E.
        pages cm
    Includes bibliographical references and index.
    ISBN 978-0-7844-1352-4 (soft cover) – ISBN 978-0-7844-7839-4 (e-book pdf)
    1. Earthquake resistant design–Standards.   2. Earthquake resistant design–
Case studies.   I. Title.
    TA658.44.C3824 2014
    624.1'7620218–dc23
                                        2014012558

Published by American Society of Civil Engineers
1801 Alexander Bell Drive
Reston, Virginia, 20191-4382
www.asce.org/bookstore | ascelibrary.org

Contents

Appendices

Preface

The purpose of this guide is to provide examples related to the use of the Standard ASCE/SEI 7-10, *Minimum Design Loads for Buildings and Other Structures* (often referred to as ASCE 7). The guide is also pertinent to users of the *2012 International Building Code* (ICC, 2011) because the IBC refers directly to ASCE 7.

Sections of ASCE 7 Pertinent to the Guide

Seismic Loads: Guide to the Seismic Load Provisions of ASCE 7-10 (the *Guide*) has examples pertinent to the following chapters of ASCE 7:

Chapter 1: General
Chapter 2: Combinations of Loads
Chapter 11: Seismic Design Criteria
Chapter 12: Seismic Design Requirements for Building Structures
Chapter 16: Seismic Response History Procedures
Chapter 20: Site Classification Procedure for Seismic Design
Chapter 22: Seismic Ground Motion and Long Period Maps

Seismic material excluded from the *Guide* are Chapter 13 (Nonstructural Components), Chapter 14 (Material-Specific Design and Detailing Requirements), Chapter 15 (Nonbuilding structures), Chapter 17 (Seismic Design Requirements for Seismically Isolated Structures), Chapter 18 (Seismic Design Requirements for Structures with Damping Systems), Chapter 19 (Soil-Structure Interaction for Seismic Design), and Chapter 21 (Site-Specific Procedures for Seismic Design).

The vast majority of the examples in the *Guide* relate to Chapters 1, 2, 11, 12, and 16 of ASCE 7, with buildings as the principal subject. The materials on nonstructural components and on nonbuilding structures will be expanded in a later edition of the *Guide*, or in a separate volume. The materials presented for Chapter 16 relate to the selection and scaling of ground motions for response history analysis and the use of linear response history analysis.

Chapter 14 of ASCE 7 is not included because the *Guide* focuses principally on seismic load analysis and not seismic design. The reader is referred to the Reference section of the *Guide* for resources containing design examples. The materials included in Chapters 17 through 19 are considered "advanced topics" and may be included in a future volume of examples.

The principal purpose of the *Guide* is to illustrate the provisions of ASCE 7 and not to provide background on the theoretical basis of the provisions. Hence, theoretical discussion is kept to a minimum. However, explanations are provided in a few instances. The reference section contains several sources for understanding the theoretical basis of the ASCE 7 seismic loading provisions. Specifically, the reader is referred to the expanded commentary to the *ASCE Seismic Provisions*. Note that this commentary was first available in the third printing of ASCE 7. Additional useful documents provided by FEMA (at no charge) are as follows:

> FEMA P-749, "Earthquake Resistant Design Concepts" (FEMA, 2010);
>
> FEMA P-750, "NEHRP Recommended Seismic Provisions for New Buildings and Other Structures" (FEMA, 2009a); and
>
> FEMA P-751, "NEHRP Recommended Provisions: Design Examples" (FEMA, 2012).

FEMA P-751 contains numerous detailed design examples that incorporate many of the requirements of ASCE 7-05 and ASCE 7-10. These examples are much more detailed than those provided in this *Guide* and concentrate on the structural design aspects of earthquake engineering, rather than just the loads and analysis side, which is the focus of the *Guide*.

The National Institute of Building Standards (NIST) provides another excellent set of seismic analysis and design references. These "technical briefs" cover various subjects, including diaphragm behavior, design of moment frames, design of braced frames, and nonlinear structural analysis. The briefs can be downloaded at no charge from www.nehrp-consultants .org.

How to Use the Guide

The *Guide* is organized into a series of individual examples. With minor exceptions, each example "stands alone" and does not depend on information provided in other examples. This means that, in some cases, information is provided in the beginning of the example that requires some substantial calculations, but these calculations are not shown. For instance, in the example on drift and P-delta effects (Example 19), the details for computing the lateral forces used in the analysis are not provided, and insufficient information is provided for the reader to back-calculate these forces. However, reference is made to other examples in the *Guide* where similar calculations (e.g., finding lateral forces) are presented. The reader should always be able to follow and reproduce all new numbers (not part of the given information) that are generated in the example.

Table and Figure Numbering

The examples presented in the *Guide* often refer to sections, equations, tables, and figures in ASCE 7. All such items are referred to directly, without specific reference to ASCE 7. For instance, a specific example might contain the statement, "The response modification factor R for the system is provided by Table 12.2-1."

References to sections, equations, tables, and figures that are unique to the *Guide* are always preceded by the letter G and use bold text. For example, the text may state that the distribution of forces along the height of the structure are listed in **Table G12-3** and illustrated in **Fig. G12-5**. In this citation, the number 12 is the example number, and the number after the dash is the sequence number of the item (that is, third table or fifth figure).

Notation and Definitions

The mathematical notation in the *Guide* follows directly the notation provided in Chapter 11 of ASCE 7. However, as the *Guide* does not use all of the symbols in ASCE 7, a separate list of symbols actually used in the *Guide* is provided in a separate section titled "Symbols Unique to the *Guide*." This list also provides definitions for new symbols that have been introduced in the *Guide*.

Computational Units

All examples in the *Guide* are developed in the U.S. customary (English) system, as follows (with the standard abbreviation in parentheses):

> Length units: inches (in.) or feet (ft)
> Force units: pounds (lb) or kips (k)
> Time units: seconds (s).

All other units (e.g., mass) are formed as combinations of the aforementioned units. A unit conversion table is provided.

Appendices and Frequently Asked Questions

In addition to the 22 individual examples, the *Guide* contains three appendices. The first appendix provides interpolation tables that simplify the process of calculating some of the values (e.g., site coefficients F_a and F_v) required by ASCE 7. The second and third appendices explain the use of web-based utilities for determining ground motion parameters and for selection of ground motion records for response history analysis.

The *Guide* also contains a special section titled "Frequently Asked Questions," where several common questions are listed, together with the author's answers. In some cases, this requires an interpretation of ASCE 7, especially when the standard is ambiguous.

User Comments

Users are requested to notify the author of any ambiguities or errors that are found in this *Guide*. Suggestions for improvement or additions are welcomed and will be included in future versions of the *Guide*.

Disclaimer

The interpretations of ASCE 7 requirements and any and all other opinions presented in this guide are those of the author and do not necessarily represent the views of the ASCE 7 Standard Committee or the American Society of Civil Engineers.

Acknowledgments

The author would like to acknowledge the following individuals for their contribution to the *Guide*:

Thomas F. Heausler, P.E., Heausler Structural Engineers, Kansas City, Kansas (reviewer for this edition of the *Guide*);

John Hooper, P.E., Magnusson Klemencic, Seattle, Washington (reviewer for a previous edition of the *Guide*);

William P. Jacobs V, P.E., Stanley D. Lindsey & Associates, LTD, Atlanta, Georgia (reviewer for a previous edition of the *Guide*);

Viral Patel, P.E., Walter P. Moore and Associates, Austin, Texas (reviewer for a previous edition of the *Guide*); and

C.J. Smith, P.E., Schnabel Engineering Associates, Blacksburg, Virginia (author of the example on site class analysis, Example 3).

Abbreviations and Symbols

Abbreviations

2D	Two-dimensional
3D	Three-dimensional
ACI	American Concrete Institute
AISC	American Institute of Steel Construction
ASCE	American Society of Civil Engineers
ASTM	Formerly American Society for Testing and Materials, now ASTM International
BRB	Buckling-restrained brace
CBF	Concentrically braced frame
CQC	Complete quadratic combination
DBE	Design basis earthquake
EBF	Eccentrically braced frame
ELF	Equivalent lateral force
FEMA	Federal Emergency Management Agency
IBC	International Building Code
LRH	Linear response history
MCE	Maximum considered earthquake
MRS	Modal response spectrum
NIST	National Institute of Standards and Technology
NGA	Next-generation attenuation
NRH	Nonlinear response history
PEER	Pacific Earthquake Engineering Research Center
RC	Reinforced concrete
SDC	Seismic Design Category
SRSS	Square root of the sum of squares

Symbols Unique to This Guide

Symbol	Definition	Introduced in Example No.
CS	Combined scale factor	6
FP	Fundamental period scale factor	6
k	Lateral stiffness of component	9
K	Structural stiffness matrix	20
M	Structural mass matrix	20
M_{pCk}	Plastic moment strength of column k	10
M_{pGk}	Plastic moment strength of girder k	10
R	Modal excitation vector	20
R_{eff}	Effective response modification coefficient	7

Symbol	Definition	*Introduced in Example No.*
S	Suite scale factor	6
S_{ai}	Spectral acceleration in mode i	20
S_{di}	Spectral displacement in mode i	20
$T_{computed}$	Period computed by structural analysis	10
T_{MF}	Period at which Eq. (12.8-5) controls C_s	7
T_{MN}	Period at which Eq. (12.8-6) controls C_s	18
V_{yi}	Story strength	10
Γ	Modal participation factor	20
δ_i	Displacement in mode i	20
Δ_o	Drift computed without P-delta effects	19
Δ_{CENTER}	Drift at geometric center of building	9
Δ_{EDGE}	Drift at edge of building	9
Δ_f	Drift computed with P-delta effects	19
ϕ	Mode shape	20
ω	Circular frequency of vibration	20

Table of Conversion Factors

U.S. customary units	International System of Units (SI)
1 inch (in.)	25.4 millimeters (mm)
1 foot (ft)	0.3048 meter (m)
1 statute mile (mi)	1.6093 kilometers (km)
1 square foot (ft^2)	0.0929 square meter (m^2)
1 cubic foot (ft^3)	0.0283 cubic meter (m^3)
1 pound (lb)	0.4536 kilogram (kg)
1 pound (force)	4.4482 newtons (N)
1 pound per square foot (lb/ft^2)	0.0479 kilonewton per square meter (kN/m^2)
1 pound per cubic foot (lb/ft^3)	16.0185 kilograms per cubic meter (kg/m^3)
1 degree Fahrenheit (°F)	1.8 degrees Celsius (°C)
1 British thermal unit (Btu)	1.0551 kilojoules (kJ)
1 degree Fahrenheit per British thermal unit (°F/Btu)	1.7061 degrees Celsius per kilojoule (°C/kJ)

Example 1

Risk Category

This example demonstrates the selection of Risk Category for a variety of buildings and other structures.

Risk Category is used in several places in ASCE 7, including
- Determination of importance factor (Section 11.5.1 and Table 1.5-2),
- Requirements for protected access for Risk Category IV structures (Section 11.5.2),
- Determination of Seismic Design Category (Section 11.6 and Tables 11.6-1 and 11.6-2), and
- Determination of drift limits (Section 12.12.1 and Table 12.12-1).

Table 1.5-1 provides general descriptions for all Risk Categories. The Risk Categories range from I (buildings and other structures that represent a low risk to human life in the event of failure) to IV (buildings and other structures that are designated as essential facilities and/or pose a substantial hazard to the community in the event of failure).

The broadness of the Risk Category descriptions in Table 1.5-1 makes the table insufficient for classifying risk in most cases. More detailed descriptions of Risk Category are provided in Table 1604.5 of the 2012 *International Building Code* (ICC, 2011). In several cases the IBC Risk Category descriptions refer to Occupancy Groups, specifically Occupancy Group I, which is used for institutional facilities. These groups, which are defined in Section 308 of the IBC, are further subdivided into four separate subgroups as follows:
- I-1: Structures or portions thereof for more than 16 persons who reside on a 24-hour basis in a supervised environment and receive custodial care. Examples include assisted living facilities and convalescent care facilities.
- I-2: Buildings and other structures used for medical care on a 24-hour basis for more than five persons who are incapable of self-preservation. Examples include hospitals and nursing homes.

- I-3: Buildings and other structures that are inhabited by more than five persons who are under restraint or security. Examples include detention centers, jails, and prisons.
- I-4: Buildings and other structures used for day care, where more than five persons of any age receive custodial care for less than 24 hours per day.

Additionally, several of the Risk Category descriptions in IBC Table 1604.5 depend on the occupant load of the building. Chapter 2 of IBC defines the occupant load as "the number of persons for which the means of egress of a building or portion thereof is designed." Occupant load is determined in accordance with Chapter 10 of the IBC.

In the following exercises, the use of Table 1.5-1 of ASCE 7 in conjunction with Table 1604.5 of the IBC is demonstrated through several scenarios. For each exercise, the structure is briefly described, the Risk Category is presented, and a discussion follows. While each exercise provides a geographic location for the building or structure under consideration, this location is not relevant to the selection of the Risk Category. These locations are provided simply to add some realism to the scenarios.

Note that selection of Risk Category can be somewhat subjective. When in doubt, the local building official should be consulted.

Exercise 1

A three-story university office and classroom building in Blacksburg, Virginia. The occupancy load is 375.

Answer

Risk Category = II.

Explanation

Risk Category II was chosen because the building has an occupancy load of fewer than 500, which is the threshold for classifying the building as Risk Category III. Note that a high school (secondary school) building with an identical configuration would have Risk Category III because the occupancy load is greater than 250.

Exercise 2

A six-story medical office building with outpatient surgical facilities located in Austin, Texas. The occupancy load for the building is 400.

Answer

Risk Category = II.

Explanation

Risk Category II applies here because the building is not open 24 hours per day and is thus not considered an IBC Occupancy Group I-2 building. Additionally, the surgical facilities are generally not used for emergencies.

| **Exercise 3** | A one-story elder-care facility (Alzheimer's care and nursing home) with an occupant load of 120, located in Savannah, Georgia. |

Answer

Risk Category = III.

Explanation

Risk Category III applies because the facility falls within IBC Occupancy Group I-2, has an occupant load of more than 50, and has no surgery or emergency care capability.

| **Exercise 4** | A 40-story casino and hotel in Reno, Nevada. Gambling rooms, ballrooms, and theaters accommodate as many as 800 people each. Total hotel occupancy load is 6,500. |

Answer

Risk Category = III.

Explanation

Risk Category III is selected because the facility has an occupancy load of more than 5,000 people.

| **Exercise 5** | Municipal courthouse and office building, containing two prisoner holding cells (a maximum of 15 prisoners in each) and a sheriff's department radio dispatcher facility, located in Richmond, California. Courtrooms have a maximum capacity of 120. |

Answer

Risk Category = III.

Explanation

Here the driving factor behind Risk Category III is the prisoner holding cells, which are in IBC Occupancy Group I-3. If the radio dispatcher facility were considered an emergency communication center, the Risk Category would go up to IV.

| **Exercise 6** | Retail fireworks building in Chattanooga, Tennessee, approximately 10,000 ft². The occupancy load is 125. |

Answer

Risk Category = II.

Explanation

Although fireworks are considered explosive, the energy released by the explosions is somewhat small (compared with, for example, a facility that stores military munitions). For this reason, Risk Category II is selected. This is a case where a discussion with the local building official would be useful because some potential exists for classification as Risk Category III.

Exercise 7

CBS News Affiliates office building in Tallahassee, Florida, that contains two studios for broadcasting local news. The facility is not a designated emergency communication center. The occupancy load is 235.

Answer

Risk Category = II.

Explanation

Risk Category II is selected because the building is not designated as an emergency communication center.

Exercise 8

A 95-story, mixed-use building in Chicago, Illinois, containing two floors of retail facilities (shops and restaurants with a maximum capacity of 60), 50 levels of office building space, and 43 levels of apartments. The building is rectangular in plan with dimensions of 150 ft by 175 ft.

Answer

Risk Category III.

Explanation

A building of this size would have an occupancy load greater than 5,000. Occupancies can be estimated for this building using Table 1004.1.2 of the 2012 IBC.

Exercise 9

One-story Greyhound bus station in Santa Fe, New Mexico. Buses enter the facility to load and unload passengers. The computed occupancy load is 350.

Answer

Risk Category = III.

Explanation

Risk Category III is selected because the occupancy load is greater than 300 and because Section 303.4 of IBC classifies transportation waiting areas as public areas of assembly.

Exercise 10   Beer manufacturing warehouse and distribution facility in Golden, Colorado. The occupancy load for this building is 125.

Answer

Risk Category = II.

Explanation

Risk Category II is used because higher categories are not appropriate. The facility cannot be designated as Risk Category I because it is not a minor storage facility.

Exercise 11   Grandstand for a college football stadium with seating for 15,000 individuals in Lubbock, Texas.

Answer

Risk Category = III.

Explanation

For this structure the occupancy load is greater than 5,000, thus requiring Risk Category III.

Exercise 12   Dockside cargo storage warehouse adjacent to the Houston Ship Channel. The building is one story with a gross floor area of 30,000 ft^2. Cargo is transported by forklifts and overhead cranes and is moved in and out daily. The cargo may contain materials (certain liquids in metal drums) considered toxic to humans. The occupancy load is 60.

Answer

Risk Category = III.

Explanation

Although the principal use of this nonessential facility is storage, significant human activity takes place in the structure, so Risk Category I is not appropriate. The storage of toxic materials generally requires Risk Category III. However, according to Section 1.5.3 of ASCE 7 (and similar language in footnote *b* of IBC Table 1604.5), the structure may be classified as Risk Category II if a hazard assessment and risk management plan can demonstrate that the release of toxic materials does not pose a threat to the public.

Exercise 13   Grain storage silo in Hays, Kansas.

Answer

Risk Category = I.

Explanation

Risk Category I here is based on the classification of this nonbuilding structure as an agricultural facility.

Exercise 14   Pedestrian bridge between an NFL football stadium and an adjacent parking lot. One end of the bridge is supported by the stadium superstructure. The bridge spans a spur of an interstate highway. The estimated maximum number of people on the bridge at any time is 220. The bridge is located in San Antonio, Texas.

Answer

Risk Category = III.

Explanation

The stadium would have a Risk Category of III because of the stadium occupancy load, which would be substantially greater than 5,000. The bridge is assigned Risk Category III because it is a means of egress from the stadium and is thereby classified as part of the stadium. (An additional consideration is the fact that a full or partial collapse of the bridge onto the interstate would inhibit the movement of emergency vehicles.)

Exercise 15   The entry foyer for a regional hospital in St. Louis, Missouri. The 1,800-ft^2, glass-enclosed structure is separate from but adjacent to the hospital. The main purpose of the foyer is for visitors to the hospital to gain access to the main hospital building. A reception desk and several unattended information kiosks are also in the foyer. A covered walkway passes between the main hospital and the foyer. Hospital staff and emergency personnel gain access through other portals.

Answer

Risk Category = II or IV.

Explanation

The hospital is clearly a Risk Category IV facility. If the foyer were an operational entry into the hospital, Section 11.5.2 of ASCE 7 would require the foyer to be classified as Risk Category IV as well. However, because hospital staff and emergency personnel do not gain access through the foyer, the entry foyer may be considered nonoperational and, hence, may be classified as a Risk Category II structure. Consultation with the local building official would be appropriate before a final designation could be made.

Example 2

Importance Factor and Seismic Design Category

This example demonstrates the determination of the seismic importance factor and the Seismic Design Category.

Importance Factor

Importance factors are a function of the Risk Category and are provided in Table 1.5-2. Values range from 1.0 for Risk Categories I and II to 1.5 for Risk Category IV. The primary use of the importance factor (I_e) is in the determination of design lateral forces. For example, Eq. (12.8-2) provides the response coefficient C_s for low period systems:

$$C_s = \frac{S_{DS}}{\left(\dfrac{R}{I_e}\right)} \qquad \text{(Eq. 12.8-2)}$$

In Eq. (12.8-2), I_e appears to be a modifier of R, which is an incorrect interpretation because R is a system-dependent parameter that is independent of risk. Another interpretation of I_e is obtained when equation Eq. (12.8-2) is written as follows:

$$C_s = \frac{S_{DS}I_e}{R} \qquad \text{(Eq. G2-1)}$$

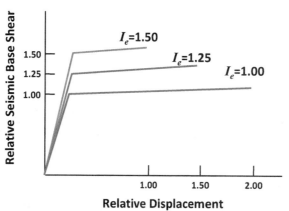

Fig. G2-1

Influence of importance factor on system performance for systems designated as "all other structures" in Table 12.12-1

In the revised equation, I_e appears to act as a multiplier of the design spectral acceleration. This also is an incorrect interpretation because the level of ground motion felt by a building is not a function of risk. The true purpose of the importance factor is to provide an *additional strength* for risk-critical facilities. For the same level of ground motion and type of detailing, a stronger building will have lower ductility demand and less damage than a weaker system.

Ductility demand is also reduced by limiting drift, and I_e (indirectly) serves this purpose as well. This can be seen by the allowable drifts provided in Table 12.12-1, which are a function of the Risk Category, and via Table 11.5-1 are also directly related to the importance factor. Thus, when considering both strength and deformation, a Risk Category IV building of a given R value, with $I_e = 1.5$, would be designed to be 1.5 times stronger than and, for most structures (see "all other structures" in Table 12.12-1), would have one-half the allowable drift of the same building designed with a Risk Category of I or II.

The comparison of system behaviors with different importance factors is shown through a set of idealized force-deformation plots in **Fig. G2-1**. The Risk Category IV building with $I_e = 1.5$ would have a significantly lower ductility demand and probably would sustain less damage than the system with $I_e = 1.0$. Damage is reduced, but not eliminated, in Risk Category IV systems.

Seismic Design Category

Seismic Design Category (SDC) is defined in Section 11.6 and Tables 11.6-1 and 11.6-2. The parameters that affect SDC are the Risk Category and the design level spectral accelerations S_{DS} and S_{D1}, or for very high level ground motions, the mapped maximum considered earthquake (MCE$_R$) (Section 11.4) spectral acceleration S_1. The SDC depends on the site class because S_{DS} and S_{D1} are directly related to the site class via Eqs. (11.4-1) and (11.4-2).

In the examples that follow, two sites, one in east Tennessee and the other near Concord, California, are considered. For each site, consideration

Determination of Seismic Design Category for Sites in East Tennessee

| Site Class | Ground Motion Parameters (g) | | | | Seismic Design Category | |
	S_S	S_1	S_{DS}	S_{D1}	*II*	*IV*
B	0.42	0.13	0.280	0.087	B	C
D	0.42	0.13	0.409	0.198	C	D

is given to the same structure constructed on soils with Site Class B or D. Consideration is given also to two different Risk Categories for each site: II and IV. The results of the calculations are presented in **Tables G2-1** and **G2-2** for the east Tennessee and Concord locations, respectively.

For the east Tennessee site, which is of relatively low seismicity, the Seismic Design Category for the Site Class B location is B for Risk Category II buildings and C for Risk Category IV buildings. For Site Class D, the Seismic Design Category is C for the Risk Category II building and D for the Risk Category IV building. The increased Seismic Design Category on Site Class D soils is caused by the site class amplification factors F_a and F_v (see subsequent calculations).

Thus, in east Tennessee, the Seismic Design Category ranges from B to D, depending on use and site. Moving from SDC B to D has important implications for the design and detailing of the structural system.

In Concord, the SDC is D in all cases. Had the ground motions been somewhat stronger, with S_1 greater than 0.75 g, the Seismic Design Category would be increased to E for Risk Category I, II, and III structures and increased to F for Risk Category IV structures.

Detailed Calculations for East Tennessee

From Figs. 22-1 and 22-2 (or from the USGS web application), $S_S = 0.42$ g and $S_1 = 0.13$ g.

Table G2-2

Determination of Seismic Design Category for Sites in Concord, California

| Site Class | Ground Motion Parameters (g) | | | | Seismic Design Category | |
	S_S	S_1	S_{DS}	S_{D1}	*II*	*IV*
B	1.74	0.60	1.16	0.40	D	D
D	1.74	0.60	1.16	0.60	D	D

For Site Class B

$$F_a = 1.0 \text{ and } F_v = 1.0 \qquad \text{(Tables 11.4-1 and 11.4-2)}$$

$$S_{DS} = \frac{2}{3} F_a S_S = \frac{2}{3}(1.0)(0.42) = 0.280 \ g \qquad \text{[Eqs. (11.4-1) and (11.4-3)]}$$

$$S_{D1} = \frac{2}{3} F_v S_1 = \frac{2}{3}(1.0)(0.13) = 0.087 \ g \qquad \text{[Eqs. (11.4-2) and (11.4-4)]}$$

Seismic Design Category = B for Risk Category II (Tables 11.6-1 and 11.6-2)

Seismic Design Category = C for Risk Category IV (Tables 11.6-1 and 11.6-2)

For Site Class D

$$F_a = 1.46 \text{ and } F_v = 2.29 \qquad \text{(Tables 11.4-1 and 11.4-2)}$$

$$S_{DS} = \frac{2}{3} F_a S_S = \frac{2}{3}(1.46)(0.42) = 0.409 \ g \qquad \text{[Eqs. (11.4-1) and (11.4-3)]}$$

$$S_{D1} = \frac{2}{3} F_v S_1 = \frac{2}{3}(2.29)(0.13) = 0.198 \ g \qquad \text{[Eqs. (11.4-2) and (11.4-4)]}$$

Seismic Design Category = C for Risk Category II (Tables 11.6-1 and 11.6-2)

Seismic Design Category = D for Risk Category IV (Tables 11.6-1 and 11.6-2)

Seismic Design Category Exception for Buildings with Short Periods

Under certain circumstances, determining the Seismic Design Category on the basis of S_{DS} only is permitted. The specific requirements are listed as four numbered points in Section 11.6. This provision applies only for systems with very short periods of vibration (with the approximate period T_a less than $0.8T_S$). This exception, where applicable, may result in the lowering of the SDC from, say, C to B, where the SDC of C would be required if the exception were not evaluated.

Example 3

Site Classification Procedure for Seismic Design

In this example, the seismic site class is determined for a given site.

Site class is used to characterize the type and properties of soils at a given site and account for their effect on the site coefficients, F_a and F_v, used in developing the design response spectrum (generalized simplified seismic analysis). The procedure can also require a site response analysis in accordance with Section 21.1, depending on the site class determination. However, the site classification procedure does not encompass evaluation of potential geologic and seismic hazards (Section 11.8). The following example is applicable for the site classification procedure provided in Chapter 20 of ASCE 7. Other codes appear similar but contain important differences. See Chapter 11 for definitions pertaining to the site classification procedure.

Based on the competency of the soil and rock material, a site is categorized as Site Class A, B, C, D, E, or F. The site classes range from hard rock to soft soil profiles as presented in **Table G3-1**. This table appears in ASCE 7 as Table 20.3-1.

For this example, the shear wave velocity criteria are not covered in detail. Shear wave velocity correlations and direct measurement require considerable experience and judgment, which are beyond the scope of this example. Proper use of shear wave velocity data requires consulting with an experienced professional.

Table G3-1   Site Classification (Table 20.3-1 of ASCE 7-10)

Site Class	$\bar{v}_s$	$\bar{N}$ or $\bar{N}_{ch}$	$\bar{s}_u$
A. Hard rock	> 5,000 ft/s	NA	NA
B. Rock	2,500 to 5,000 ft/s	NA	NA
C. Very dense soil and soft rock	1,200 to 2,500 ft/s	> 50	> 2,000 lb/ft^2
D. Stiff soil	600 to 1,200 ft/s	15 to 50	1,000 to 2,000 lb/ft^2
E. Soft clay soil	< 600 ft/s	< 15	< 1,000 lb/ft^2
	Any profile with more than 10 ft of soil having the following characteristics: - Plasticity index PI > 20, - Moisture content $w \geq 40\%$, and - Undrained shear strength $\bar{s}_u < 500$ lb/ft^2		
F. Soils requiring site response analysis in accordance with Section 21.1	See Section 20.3.1		

Data Collection

To classify a site, the proper subsurface profile and necessary data need to be obtained. According to Section 20.1:

- Site soil shall be classified based on the upper 100 ft (30 m) of the site profile.
- In absence of data to a depth of 100 ft, soil properties may be estimated by the registered design professional preparing the soil investigation.
- Where soil properties are not known in sufficient detail, Site Class D shall be used unless Site Class E or F soils are determined to be present.

Where site class criteria are based on soil properties (PI, w, s_u), the values are to be determined by laboratory tests as specified: Atterberg limits (ASTM D4318, 2005b), moisture content (ASTM D2216, 2005a), and undrained shear strength (ASTM D2166, 2006, or ASTM D2850, 2007).

Site Class Determination

The procedure can be generally broken into three steps as follows. **Fig. G3-1** summarizes the steps.

Fig. G3-1

Site class determination
flowchart

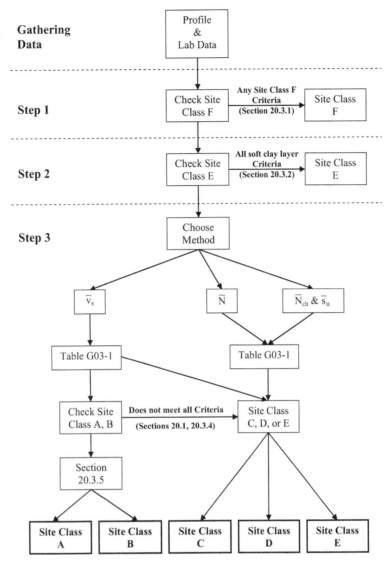

Step 1: Check Site Class F (Section 20.3.1)

If *any* of the following conditions are met, the site shall be classified as Site Class F:

1. Soils vulnerable to potential failure or collapse under seismic loading, such as
   - liquefiable soils,
   - quick or highly sensitive clays, and
   - collapsible or weakly cemented soils.

   Exception: $T \leq 0.5$ s (Section 20.3.1)

   Or
2. Peat and/or highly organic clays [$H > 10$ ft (3 m)].

   Or
3. Very high plasticity clays [$H > 25$ ft (7.6 m) with PI > 75].

   Or
4. Very thick soft to medium stiff clays [$H > 120$ ft (37 m)] with s_u < 1,000 lb/ft^2 (50 kPa).

A site response analysis (Section 21.1) shall be performed for sites determined to be Site Class F.

Step 2: Check Site Class E (Section 20.3.2)

If a profile contains a soft clay layer with *all* of the following characteristics, the site shall be classified as Site Class E:
1. $H_{\text{layer}} > 10$ ft (3 m),
2. PI > 20,
3. $w \geq 40\%$, and
4. $s_u < 500$ lb/ft^2 (25 kPa).

Step 3: Check Site Class A, B (Sections 20.3.4 and 20.3.5) or Site Class C, D, E (Section 20.3.3)

Using one of the following three methods, categorize the site using **Table G3-1**. All computations of $\bar{v}_s$, $\bar{N}$, $\bar{N}_{ch}$, and $\bar{s}_u$ shall be performed in accordance with Section 20.4.

$\bar{v}_s$ Method

An advantage of using shear wave velocity data is that the measured behavior better characterizes the subsurface profile than data collected from point locations (i.e., borings). Disadvantages of the method include the somewhat high cost and experience required to perform the method and interpret the data.

If appropriate shear wave velocity data are available, $\bar{v}_s$ shall be calculated for the top 100 ft (30 m) using Eq. (20.4-1) and the appropriate site class determined from **Table G3-1**.

If the classification falls into criteria of either Site Class A or B in **Table G3-1**, the following additional criteria shall be considered:
- Site Class A or B shall not be assigned to a site if more than 10 ft (3 m) of soil lies between the rock surface and the bottom of the spread footing or mat foundation (Section 20.1).
- Shear wave velocity criteria specified in Section 20.3.4 for Site Class B and in Section 20.3.5 for Site Class A shall be observed.

The applicable site class depends on which of the aforementioned criteria are met.

$\bar{N}$ Method

Using standard field penetration values for all soil and rock layers, $\bar{N}$ shall be calculated for the top 100 ft (30 m) using Eq. (20.4-2). The following should be considered regarding standard field penetration values (Section 20.4.2):
- ASCE 7 states that standard penetration resistance values as "directly measured in the field without corrections" should be used. The author believes that energy corrections based on the type of hammer used should be applied because this difference is fundamental in the values measured. For instance, the standard penetration values from an automatic hammer should be appropriately

reduced to safety hammer (N_{60}) values to reflect the high efficiency of the automatic hammer.

- Eq. (20.4-2) requires a single N value for each distinct layer in the profile. Average or conservatively choose values from multiple borings to characterize each distinct layer.
- Use a maximum of 100 blows/ft. See Section 20.4.2 for discussion.
- Where refusal is met for a rock layer, N shall be taken as 100 blows/ft.

$\bar{N}_{ch}$ and $\bar{s}_u$ Method

Divide the 100-ft (30-m) profile into cohesionless and cohesive layers in accordance with the definitions presented in Section 20.3.3.

Using standard field penetration values for the cohesionless layers, $\bar{N}_{ch}$ shall be calculated using Eq. (20.4-3). The comments on standard field penetration values listed in the previous section entitled "$\bar{N}$ Method" apply to $\bar{N}_{ch}$.

Using undrained shear strength values for the cohesive layers, $\bar{s}_u$ shall be calculated using Eq. (20.4-4). As stated in Section 20.4.3, undrained shear strength values shall be determined in accordance with ASTM D2166 or ASTM D2850.

Determining a site class requires two steps, first using $\bar{N}_{ch}$ as the classification criterion, and second using $\bar{s}_u$ as the classification criterion, using **Table G3-1**. If the site classes differ, the site shall be assigned a site class corresponding to the softer soil (Section 20.3.3).

Site Classification Example

The site profile presented in this example represents highly idealized subsurface conditions. Interpretation of actual subsurface data and soil properties requires substantial judgment by the geotechnical professional. The site profile used in the example is shown in **Fig. G3-2**.

The blow counts in the example represent N_{60} values obtained from a safety hammer. As noted in the $\bar{N}$ Method section, the author believes these are the appropriate values to be used in seismic site class determination. The steps previously outlined are applied to the given example below.

Step 1: Check Site Class F

If the profile meets *any* of the criteria in Section 20.3.1, the site shall be classified as Site Class F. This profile has been chosen to ensure that Site Class F does not apply. However, this check should not be overlooked in practice.

Step 2: Check Site Class E

If the profile contains any layers meeting *all* criteria in Section 20.3.2, the site shall be classified as Site Class E. Soft clay layer criteria are checked below (bold text indicates a criterion that is not met):

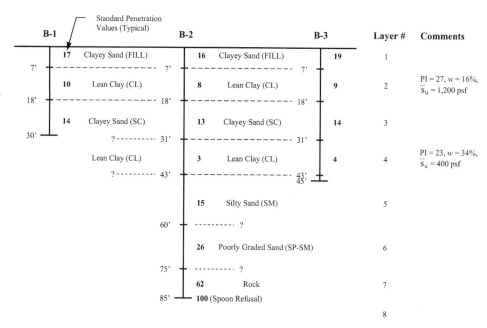

Fig. G3-2

Subsurface profile

Notes: Standard penetration values presented as N_{60} values (ASTM 1586). Soil designations are based on USCS classification (ASTM 2487).

Soft clay:	$H > 10$ ft	PI > 20	$w \geq 40\%$	$\bar{s}_u < 500$ lb/ft^2
Layer 2:	$H = 11$ ft	PI = 27	**$w = 16\%$**	**$\bar{s}_u = 1,200$ lb/ft^2**
Layer 4:	$H = 12$ ft	PI = 23	**$w = 34\%$**	$\bar{s}_u = 400$ lb/ft^2

Layer 2 does not qualify based on its water content and undrained shear strength. Layer 4 does not qualify based on its water content.

Because neither layer satisfies *all* soft clay layer criteria, the site does not automatically qualify for Site Class E.

Step 3: Check Site Class A and B or Site Class C, D, E

For this step, the $\bar{N}$ method determines site class. Using this method automatically excludes Site Class A or B because they are based on shear wave velocity.

Using the notation from Eq. (20.4-2) and the example profile in **Table G3-2**, the site can be classified as follows: the value of $\bar{N} = 100/8.65 = 12$ calculated using Eq. (20.4-2) classifies the site as Site Class E ($\bar{N} < 15$).

Some observations that can be made from the values in **Table G3-2** are the following:

- Standard penetration value of 50 blows/in. at refusal (85 ft) was assigned a maximum value allowed of 100 blows/ft (Section 20.4.2).
- Based on the known geology, this blow count was then used from refusal to a depth of 100 ft to complete the site profile, resulting in a 15-ft layer with blow counts of 100 blows/ft.

Table G3-2 Summary of $\bar{N}$ Method

Layer No. (i)	Soil or Rock Designation	Cohesionless[a]	Cohesive[a]	N_i (blows/ft)	d_i (ft)	d_i/N_i
1	SC (FILL)	X		17	7	0.41
2	CL		X	9	11	1.22
3	SC	X		13	13	1.00
4	CL		X	3	12	4.00
5	SM	X		15	17	1.13
6	SP-SM	X		26	15	0.58
7	Rock	X		62	10	0.16
8	Rock	X		100	15	0.15
					Total 100	Total 8.65
					$\bar{N} = 100/8.65 = 12$	

[a]Based on ASCE 7 definition, Section 20.3.3.

Comments on Site Classification

Although highly idealistic, the site profile used in this example illustrates the need for adequate site investigation. Had only the data from shallow boring (B-1) been available, the designer would be unaware of the potentially soft clay layer encountered in B-2 and B-3.

Other important considerations that are not explicitly covered by ASCE 7 include

- Where to begin the site profile for below-grade structures;
- How to incorporate planned site grading (cut and fill) at the site;
- How to apply the design response spectrum method (generalized simplified seismic analysis) to structures supported on deep foundations; and
- How to characterize highly variable site profiles (e.g., layer thickness and/or properties) within a given site.

Example 4

Determining Ground Motion Parameters

In this example, the design basis spectral accelerations S_{DS} and S_{D1} are found for a site in Savannah, Georgia. They are first approximately determined by hand, using maps and tables provided by ASCE 7, and are then checked with a software utility provided by the U.S. Geological Survey.

The basic ground motion parameters in ASCE 7 are S_S and S_1. S_S is the "short period" spectral acceleration ($T = 0.2$ s), and S_1 is the one-second ($T = 1.0$ s) spectral acceleration for sites on firm rock (Site Class B). These accelerations are based on the risk-based maximum considered earthquake (MCE_R), for which approximately a 2.0% probability exists of being exceeded in 50 years.

S_S and S_1 are used in several ways in ASCE 7; the most important is in the determination of the design-level acceleration parameters S_{DS} and S_{D1}. The design accelerations include a site coefficient factor (F_a or F_v) that accounts for soil characteristics different from firm rock and a multiplier of 2/3, which effectively converts from the MCE_R basis to a somewhat lower level of shaking, called the design basis earthquake (DBE). The site coefficients are obtained by interpolation from values provided in Tables 11.4-1 and 11.4-2.

S_S and S_1 are obtained from maps (Figs. 22-1 and 22-2) in Chapter 22. Due to the low resolution of the maps, the S_S and S_1 values obtained from the maps may be very approximate. Thus, determining the values of S_S and S_1 from a web-based computer program provided by the U.S. Geological Survey (USGS) is more common. This program can also provide values for the site coefficients F_a and F_v. This example illustrates the use of the maps and tables and is then reworked using the USGS program.

Example: Find Ground Motion Values for a Site in Savannah, Georgia

This example considers a site on Site Class D soil in downtown Savannah, Georgia. Savannah lies on the Atlantic coast, just south of the border between South Carolina and Georgia. **Fig. G4-1** shows the building site location with a small star. The broken line in the ASCE 7 contour maps represents the border between the states. The small cross in **Fig. G4-2(a)** and **Fig. G4-2(b)**, which are taken directly from Figures 22-1 and 22-1, shows the site location on the ASCE 7 contour maps.

    Fig. G4-2(a) indicates that Savannah lies close to the 30% gravity (0.3 g) contour, so an acceleration of 0.3 g is used for S_S. **Fig. G4-2(b)** indicates that Savannah is somewhat closer to the 10% gravity (0.1 g) contour

Fig. G4-1

Building site in Savannah, Georgia

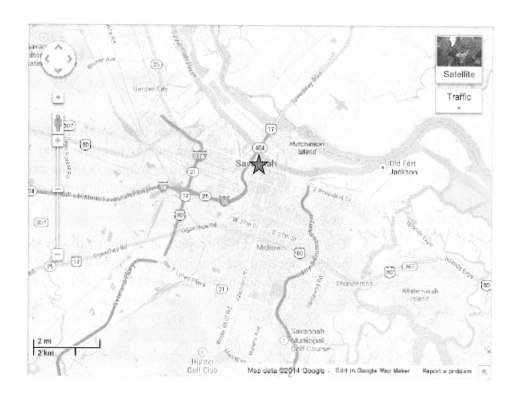

Fig. G4-2

Spectral acceleration contours for S_s and S_1 in Savannah, Georgia

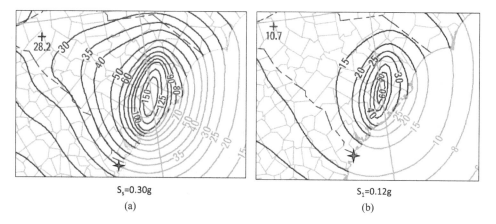

$S_s = 0.30g$

(a)

$S_1 = 0.12g$

(b)

Fig. G4-3

Interpolating for site coefficients F_a and F_v

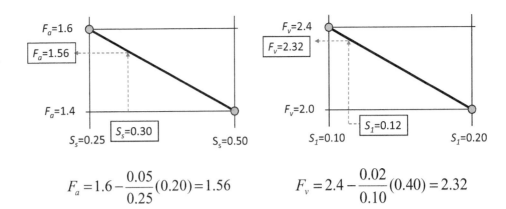

$$F_a = 1.6 - \frac{0.05}{0.25}(0.20) = 1.56 \qquad F_v = 2.4 - \frac{0.02}{0.10}(0.40) = 2.32$$

than it is to the 15% gravity (0.15 g) contour, so a value of 0.12 g will be used for S_1. Summarizing, the following values are used for Savannah:

$$S_S = 0.30 \ g$$
$$S_1 = 0.12 \ g$$

The site coefficients F_a and F_v are taken from Tables 11.4-1 and 11.4-2, respectively. Table 11.4-1 shows that for Site Class D it will be necessary to interpolate between values of $F_a = 1.6$ for $S_S = 0.25$ g and $F_a = 1.4$ for $S_S = 0.50$ g. **Fig. G4-3(a)** indicates that $F_a = 1.56$. Interpolation is also required to determine F_v. **Fig. G4-3(b)** shows that $F_v = 2.32$. Using these site coefficients, the site-amplified ground motion parameters are computed as follows:

$$S_{MS} = F_a S_S = 1.56(0.30) = 0.468 \ g \qquad \text{(Eq. 11.4-1)}$$

$$S_{M1} = F_v S_1 = 2.32(0.12) = 0.278 \ g \qquad \text{(Eq. 11.4-2)}$$

Although interpolating from Tables 11.4-1 and 11.4-2 is not difficult, it is somewhat inconvenient. For this reason, a variant of the tables is provided in **Appendix A** of this book in **Tables GA-1** and **GA-2**, in which the coefficients in the original tables are replaced by interpolation formulas. The example is reworked as follows:

from **Fig. GA-1** and **Table GA-1**:

$$F_a = 1.8 - 0.8 \ S_S = 1.8 - 0.8(0.3) = 1.56$$

and from **Fig. GA-2** and **Table GA-2**:

$$F_v = 2.8 - 4.0 \ S_1 = 2.8 - 4.0(0.12) = 2.32$$

These are the same values determined from interpolation.

Given the site-amplified ground motion values, the design ground motion values are obtained as follows:

$$S_{DS} = (2/3)S_{MS} = (2/3)(0.468) = 0.312 \ g \qquad \text{(Eq. 11.4-3)}$$

$$S_{D1} = (2/3)S_{M1} = (2/3)(0.278) = 0.185 \ g \qquad \text{(Eq. 11.4-4)}$$

It is important to note that in these equations, S_{DS} and S_{D1} are expressed in "g" units. Thus, if $g = 386$ in./s^2, $S_{DS} = 0.312 \times 386 = 120.4$ in./s^2. This is consistent with the concept that these terms represent spectral acceleration. When the terms S_{DS} and S_{D1} are used to determine the design base shear coefficient, C_s, in Eq. (12.8-2) (for example), they are referred to as "acceleration parameters," and in this context, no units should be assigned to the terms. This inconsistent use of units for the same term should be addressed in future versions of ASCE 7.

Use of the USGS Ground Motion Calculator

Due to the lack of detailed contour maps in ASCE 7-10, obtaining the spectral ordinates S_S and S_1 from the "Design Maps" web application maintained by the USGS is usually more convenient. The web page containing the calculator has the following address: http://earthquake.usgs.gov/designmaps. **Appendix B** of this guide has detailed information regarding the USGS utility.

The main screen for the design maps utility is shown in **Fig. G4-4**. Data entered into the utility include the design code reference document (ASCE 7-10), the report title, the site class, the Risk Category, and the site's latitude and longitude. When using latitude and longitude, it is important to note that longitude values must be entered as a negative number (because the site is west of the Prime Meridian).

Fig. G4-4

Main screen from the USGS ground motion calculator

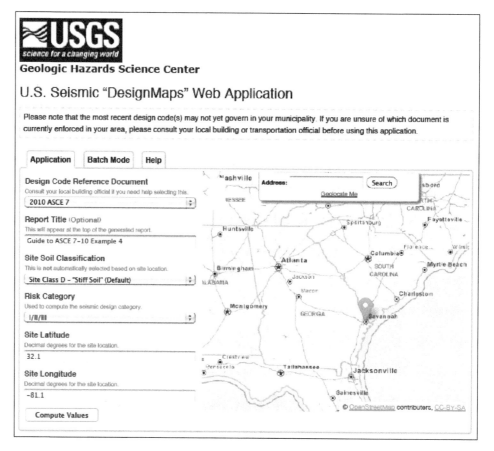

Fig. G4-5

Ground motion parameters from USGS ground motion calculator

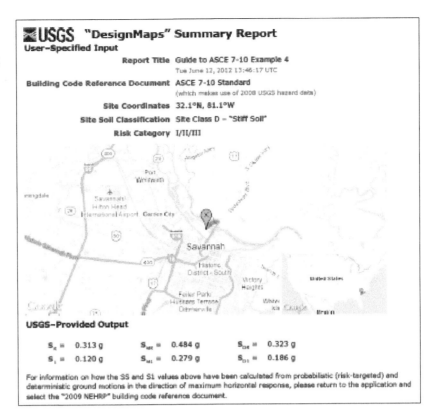

The USGS utility also has the capability to calculate the site class factors F_a and F_v and then the resulting design acceleration values S_{DS} and S_{D1}. Plots of the design response spectra (Fig. 11.4-1) in a variety of formats are also available through the utility.

The use of the USGS ground motion calculator is illustrated for the aforementioned example in downtown Savannah, Georgia. The latitude and longitude ordinates for Site Class D site, near the intersection of Anderson Street and East Broad Street are as follows:

$$\text{latitude} = 32.06 \text{ deg}$$
$$\text{longitude} = -81.09 \text{ deg}$$

The calculator (see **Fig. G4-5**) provides the following values:

S_S	= 0.313 g	S_1	= 0.120 g
S_{MS}	= 0.484 g	S_{M1}	= 0.279 g
S_{DS}	= 0.323 g	S_{D1}	= 0.186 g
F_a	= 1.55	F_v	= 2.32 (obtained from these values)

These are close to the values computed by hand. This result is expected. The calculator is generally much easier to use than is the hand method and is less prone to error, so it is generally preferred.

Example 5

Developing an Elastic Response Spectrum

In this example, an elastic response spectrum is generated for a site in Savannah, Georgia. An elastic spectrum is used directly in the modal response spectrum analysis approach (Section 12.9) and as a basis for ground motion scaling in response history analysis (Chapter 16).

This example will use the same location in Savannah, Georgia, that was the setting for Example 4. This site is in downtown Savannah and is Site Class D. The basic ground motion parameters, obtained through the use of the USGS ground motion calculator, are

$$
\begin{array}{llll}
S_S & = 0.313\ g & S_1 & = 0.120\ g \\
F_a & = 1.55 & F_v & = 2.32 \\
S_{MS} & = 0.484\ g & S_{M1} & = 0.279\ g \\
S_{DS} & = 0.323\ g & S_{D1} & = 0.186\ g
\end{array}
$$

The basic form of the design response spectrum is shown in Fig. 11.4-1 of ASCE 7. This spectrum has four branches:

1. A straight-line ascending portion between $T = 0$ and $T = T_0$ [Eq. (11.4-5)],
2. A constant acceleration portion between $T = T_0$ and $T = T_S$ ($S_a = S_{DS}$),
3. A descending "constant velocity" region between $T = T_S$ and $T = T_L$ [Eq. (11.4-6)], and

4. A descending "constant displacement" region beyond T_L [Eq. (11.4-7)].

The four branches of the spectrum are controlled by the design-level spectral accelerations S_{DS} and S_{D1}; by Eq. (11.4-5), which describes the first branch of the spectrum; and by the "long-period transition period" T_L, which is provided by the contour maps of Figs. 22-12 through 22-16. For Savannah, the long-period transition, obtained from Fig. 22-12, is equal to 8 s. Such a long period would apply only for tall or very flexible buildings or for the sloshing of fluid in tanks.

The other transitional periods, T_0 and T_S, are computed according to Section 11.4.5 as follows:

$$T_S = S_{D1}/S_{DS} = 0.186/0.323 = 0.576 \text{ s}$$

$$T_0 = 0.2T_S = 0.115 \text{ s}$$

The spectral acceleration at $T = 0$ is given by Eq. (11.4-5). When $T = 0$, this equation produces an acceleration of $0.4(S_{DS}) = 0.4(0.323) = 0.129$ g. This result is an approximation of the design-level peak ground acceleration. More accurate values of peak ground acceleration can be found in Figures 22-7 to 22-11.

The complete response spectrum for the Site Class D location in Savannah is plotted with a bold line in **Fig. G5-1**. The spectrum is plotted for a maximum period of 4.0 s, and as such, the fourth "constant displacement" branch is not shown.

For use in a computer program, the response spectrum is often presented in a table of period-acceleration values. In some cases, the spectrum is automatically generated from values of S_{DS} and S_{D1}.

Fig. G5-1

Elastic design response spectra ($R = 1$, $I_e = 1$) for various site classes where $S_S = 0.313$ g and $S_1 = 0.120$ g

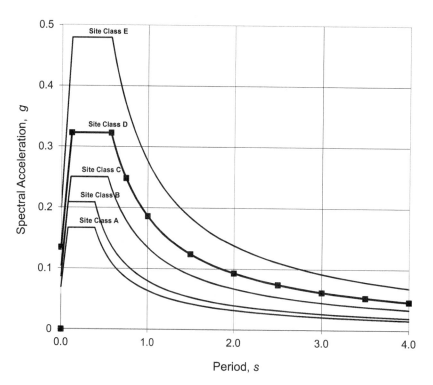

Table G5-1   Elastic Design Spectral Ordinates ($R = 1$, $I_e = 1$) for a Site Class D Location in Savannah, Georgia

Period T, s	Spectral Acceleration S_a (g)
0	0.134
0.115 (T_0)	0.323
0.576 (T_s)	0.323
0.65	0.286
0.75	0.248
1.00	0.186
1.50	0.124
2.00	0.093
2.50	0.074
3.00	0.062
3.50	0.053
4.00	0.047

Table G5-2   Elastic Design Response Spectrum Parameters ($R = 1$, $I_e = 1$) for Various Site Classes where $S_S = 0.313\ g$, $S_1 = 0.120\ g$

Site Class	F_a	F_v	S_{DS} (g)	S_{D1} (g)
A	0.80	0.80	0.167	0.064
B	1.00	1.00	0.209	0.080
C	1.20	1.68	0.251	0.135
D	1.55	2.32	0.323	0.186
E	2.30	3.44	0.479	0.276

When providing a table of spectrum values, it is important to provide sufficient resolution in the curved portions. Discrete spectral values are provided in **Table G5-1** for the Savannah site. These points are also represented by square symbols on the Site Class D spectrum in **Fig. G5-1**.

It is important to recognize that the elastic response spectrum developed in this example has not been adjusted by the importance factor I_e, nor by the response modification coefficient R. The use of these parameters and the deflection amplification parameter C_d, are described in Section 12.9.2 of ASCE 7.

For illustrative purposes only, response spectra are also shown in **Fig. G5-1** for Site Classes A, B, C, and E. Parameters used to draw the curves are shown in **Table G5-2**. **Fig. G5-1** shows that site class can have a profound effect on the level of ground acceleration for which a structure must be designed. For the example given, S_{DS} for Site Class D is 1.55 times the value when the site class is B. S_{D1} increases by a factor of 2.32 when the site class changes from B to D.

Example 6

Ground Motion Scaling for Response History Analysis

Chapter 16 of ASCE 7 provides the requirements for performing a linear or a nonlinear response history analysis. Among the key components of such an analysis are the selection of an appropriate suite of ground motions and the scaling of these motions. This example emphasizes the scaling procedure and provides only minimal background on the ground motion selection process. The scaling procedures are demonstrated for both two-dimensional (2D) and three-dimensional (3D) analysis. The scaling for 2D analysis is utilized in Example 21, wherein a modal response history analysis is performed for a simple moment-resisting frame.

The scaling procedure is applied to a building in Savannah, Georgia. The building is in Seismic Design Category C and is situated on Site Class D soils. The site is not within 10 km of any known fault, so only far-field ground motions are considered.

The design-level spectral accelerations (see Section 11.4.4) are as follows:

$$S_{DS} = 0.323 \ g$$
$$S_{D1} = 0.186 \ g$$

The scaling procedure is demonstrated for both 2D and 3D analysis.

Selection of Ground Motions

Sections 16.1.3.1 and 16.1.3.2 cover ground motion selection and scaling for 2D and 3D analysis, respectively. In both cases, the ground motions must be selected from actual records, and they must have magnitude, fault distance, and source mechanism consistent with those that control the maximum considered earthquake.

Various sources of recorded ground motions exist. For this example, the next-generation attenuation (NGA) record set, provided by the Pacific Earthquake Engineering Research Center (PEER) was used. This record set is an updated version of the PEER Strong Motion Database. The NGA records are available from http://peer.berkeley.edu/nga. The website provides a search engine that allows the user to find ground motions by name (e.g., Northridge), magnitude range, and distance. Each ground motion record set consists of two horizontal records and one vertical acceleration record. The vertical record is generally not used for analysis. The PEER NGA database contained more than 3,500 record sets when this example was prepared. Appendix C of this guide provides additional information on the use of the PEER NGA website and database.

Section 16.1.4 of ASCE 7 requires that at least three record sets be used in any analysis. If fewer than seven sets are used, the response parameters used for design are the maximum values obtained from all of the analyses. If seven or more records are used, the design may be based on the average values obtained from the analysis. Because of this requirement, using seven or more ground motions is likely to be highly beneficial. To maintain simplicity in this example, however, only three ground motion sets are used.

The ground motions selected for the analysis are listed in **Table G6-1**, and ground motion parameters are presented in **Table G6-2**. Each of these motions is considered to be a far-field motion because the epicentral distance

Table G6-1 Record Sets Used for Analysis

Earthquake	PEER NGA ID	Year	Magnitude	ASCE 7 Site Class	Fault Type	Epicentral Distance (km)
San Fernando	68	1971	6.6	D	Thrust	39.5
Imperial Valley	169	1979	6.5	D	Strike-slip	33.7
Northridge	953	1994	6.7	D	Thrust	13.3

Table G6-2 Record Set Maxima

	Component 1			Component 2		
Earthquake	Bearing (deg)	PGA (g)	PGV (in./s)	Bearing (deg)	PGA (g)	PGV (in./s)
San Fernando	090	0.210	7.45	180	0.174	5.85
Imperial Valley	262	0.238	10.23	352	0.351	13.00
Northridge	009	0.416	23.2	279	0.516	24.71

Fig. G6-1

Pseudoacceleration spectra for selected ground motions

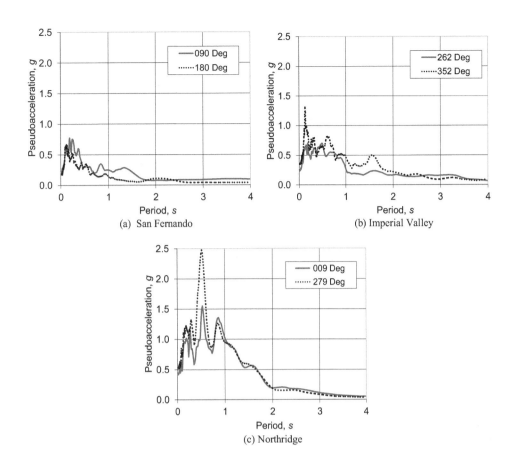

(a) San Fernando

(b) Imperial Valley

(c) Northridge

is greater than 10 km. Each of the motions was recorded in Site Class D soil, and source mechanisms (fault type) are as shown in **Table G6-1.** The magnitudes of the earthquakes, in the range of 6.5 to 6.7, are somewhat lower than that which could be expected for the maximum considered earthquake (MCE). Few MCE-level records are available, however, so these records have to suffice.

Five percent damped pseudoacceleration response spectra for the two horizontal components of motions of each earthquake are shown in **Figs. G6-1(a)** through **(c).** The spectra were generated using the *NONLIN* program (Charney et al., 2010). *NONLIN* can read the PEER database files if they are saved to a text file with an "nga" file extension, for example, northridge .nga. It is clear from these spectra that the Northridge ground motion record is dominant.

Scaling for 2D Analysis

For 2D analysis, the "strongest" components from each ground motion pair, in terms of peak ground acceleration, are used. These components are as follows:

- San Fernando 090,
- Imperial Valley 352, and
- Northridge 279.

Fig. G6-2

Various spectra resulting from the 2D ground motion scaling process

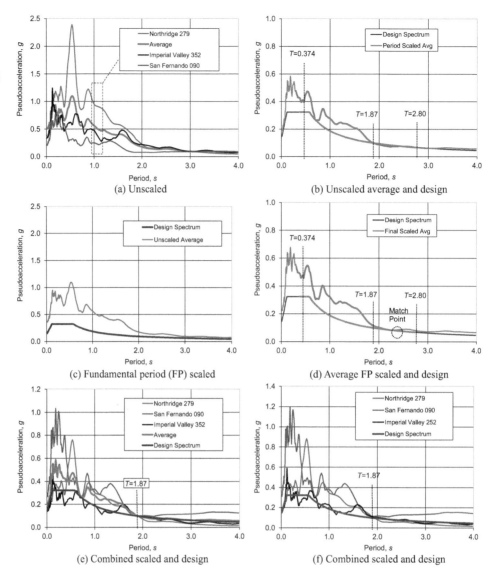

(a) Unscaled

(b) Unscaled average and design

(c) Fundamental period (FP) scaled

(d) Average FP scaled and design

(e) Combined scaled and design

(f) Combined scaled and design

Other choices are possible for selecting the component to use, such as peak ground velocity, pseudoacceleration at the structure's fundamental period, spectral shape, or the analyst's judgment and experience.

The pseudoacceleration response spectra and the average of the spectra for the strongest components are shown in **Fig. G6-2(a)**. Section 16.1.3.1 requires that the ground motions "be scaled such that the average value of the 5% damped response spectra for the suite of motions is not less than the design response spectrum for the site for periods ranging from 0.2*T* to 1.5*T*." The period would generally be determined from the same analysis model that will be eventually subjected to the response history analysis.

Given that each ground motion has its own scale factor, there are an infinite number of ways to scale the suite of motions such that the ASCE 7 scaling requirements are met. In this example, a two-step scaling approach is used, which has the advantage of producing a unique set of scale factors for a given ground motion record set.

Step 1

Scale each ground motion such that it has the same spectral acceleration as the design spectrum at the structure's fundamental period of vibration. This step results in a different scale factor, FP_i, for each motion i, wherein the abbreviation FP refers to the fundamental period scaled motions.

Step 2

A second scale factor, called S for suite scale factor, is applied to each of the FP motions, such that the ASCE 7 scaling requirement is met. The combined scale factor, C_i for each motion i, is S times FP_i. The procedure is applied to the moment frames structure of Examples 20 and 21, which has a period of vibration of 1.87 s.

Fig. G6-2(a) shows the unscaled spectra for each ground motion and the average spectra. The number shown in the legend after the ground motion (e.g., 279 after Northridge) is the compass bearing of the ground motion. The dashed box at $T = 1.0$ s correlates the ground spectra shown in the plot with the ground motion identifier in the legend. For example, the Northridge earthquake is at the top of the legend and is the topmost curve in the dashed box. The unscaled average spectrum and the design spectrum (see Section 11.4.5 of ASCE 7) are shown in Fig. G6-2(b). It is clear from this plot that the average ground motion spectrum falls above the design spectrum at all periods.

The spectral ordinates at $T = 1.87$ s are shown below for the design spectrum and for each of the individual ground motion spectra. The FP factor for each ground motion is also shown. When these scale factors are applied to the ground motions, the resulting spectra have a common ordinate of 0.0994 g at $T = 1.87$ s. This result is shown in Fig. G6-2(c). The average of the FP spectra are shown together with the design spectrum in Fig. G6-2(d).

Design spectrum: 0.0994 g

San Fernando: 0.0745 g $FP_1 = 0.0994/0.0745 = 1.333$

Imperial Valley: 0.249 g $FP_2 = 0.0994/0.249 = 0.399$

Northridge: 0.326 g $FP_3 = 0.0994/0.326 = 0.305$

The S scale factor, where S refers to spectrum, is determined such that no spectral ordinate in the combined average scaled ground motion spectrum falls below the design spectrum in the period range of 0.2 T to 1.5 T. With $T = 1.87$ s, this range is 0.374 to 2.81 s. The required S scale factor is 1.181. As seen from Fig. G6-2(d), the match point (inside the small dashed circle) occurs at a period of about 2.3 s. The final combined scale factors for each ground motion are as follows:

San Fernando: $C_1 = 1.333 \times 1.181 = 1.575$,

Imperial Valley: $C_2 = 0.399 \times 1.181 = 0.471$, and

Northridge: $C_3 = 0.305 \times 1.181 = 0.360$

Fig. G6-2(f) shows the C scaled ground motions together with the design spectrum. The ground motion ordinates at the structure's fundamental period of vibration, 1.87 s, are slightly above the ordinate for the target spectrum, but lower period ordinates of the scales spectra are significantly greater that those of the target. It appears, therefore, that the scaling procedure has resulted in individual ground motions that are much stronger in the higher modes than that which would be implied by the design spectrum.

Scaling for 3D Analysis

The ground motion scaling requirements for 3D analysis are provided in Section 16.1.3.2. The scaling procedures are similar to those for 2D analysis, with the following exceptions:

1. For each earthquake in the suite, the square root of the sum of squares (SRSS) of the spectra for each pair of horizontal components is computed. When computing the SRSS, the motion as recorded without scale factors is used.

2. Individual scale factors are applied to the SRSS spectra such that the average of the scaled SRSS spectra does not fall below 1.0 times the design for any period between $0.2T$ and $1.5T$.

With regard to point 2, the period T is generally different in the two orthogonal directions, and thus the scale factors would be different in the two directions.[1] Selection of the two periods for 3D analysis may not be straightforward for buildings in which lateral and torsional response are strongly coupled.

The scaling procedure is illustrated as follows for a building with a period of vibration of 1.5 s. As with the 2D approach, two scale factors are determined for each earthquake: a fundamental period scale factor FP_i, which is unique for each earthquake, and a suite scale factor, S, which is common to all earthquakes. The product of these two scale factors is the combined scale factor, S_i.

For this example, the structure is located in southern California on Site Class D soils. The period of vibration is taken as 1.5 seconds in each direction. The design-level spectral accelerations (see Section 11.4.4) are as follows:

$$S_{DS} = 0.323 \ g, \text{ and}$$
$$S_{D1} = 0.186 \ g.$$

The SRSS of the ground motion pairs and the average of the SRSS are shown in **Fig. G6-3(a)**. The average of the SRSS spectra is shown, together

---

[1] Having different scale factors in the two different directions is not rational, and a different interpretation of the ASCE 7 requirements is warranted. One approach for handling different periods in different direction is to select the scaling range as $0.2T_{small}$ to $1.5 \ T_{large}$, where T_{small} and T_{large} are the smaller and larger of the two first fundamental periods of vibration. Another approach would be to select the scaling range as $0.2T_{avg}$ to $1.5T_{avg}$, where T_{avg} is the average of the two periods.

Fig. G6-3

Various spectra resulting from the 3D ground motion scaling process

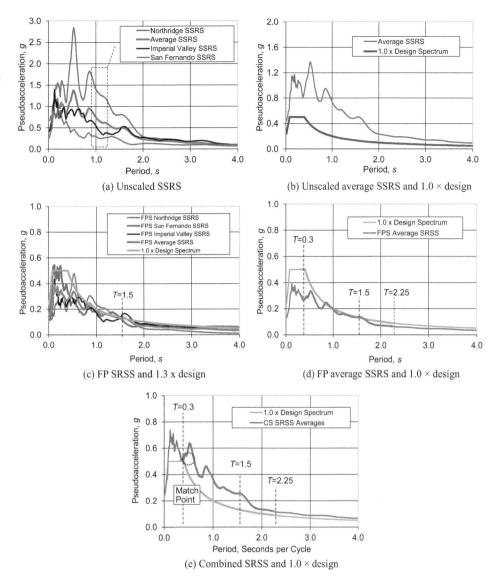

(a) Unscaled SSRS

(b) Unscaled average SSRS and 1.0 × design

(c) FP SRSS and 1.3 × design

(d) FP average SSRS and 1.0 × design

(e) Combined SRSS and 1.0 × design

with 1.0 times the design spectrum in **Fig. G6-3(b)**. The first step in the scaling process is to scale each SRSS such that the spectral acceleration at the structure's fundamental period matches 1.0 times the design spectrum at the same period.

The spectral ordinate at $T = 1.87$ seconds is 0.134 g. The appropriate scale factors were determined as follows:

> 1.0 × design spectrum: 0.134 g,
> San Fernando SRSS: 0.223 g $FP_1 = 0.134/0.223 = 0.600$,
> Imperial Valley SRSS: 0.428 g $FP_2 = 0.134/0.428 = 0.313$, and
> Northridge SRSS: 0.797 g $FP_3 = 0.134/0.797 = 0.168$.

The FS spectra are shown with 1.0 times the design spectrum in **Fig. G6-3(c)**, and the average of the FS spectra is shown with the design spectrum in **Fig. G6-3(d)**. **Fig. G6-3(d)** also shows the period range over which the S scale factor is to be determined.

The S scale factor is computed such that no spectral ordinate in the combined average scaled SRSS ground motion spectrum falls below 1.0 times the design spectrum in the period range of $0.2T$ to $1.5T$. The required S scale factor is 1.896. As seen from **Fig. G6-3(e)**, the controlling match point occurs at a period of approximately 0.36 s. The final scale factors for each ground motion are as follows:

$$\text{San Fernando SRSS: } C_1 = 0.600 \times 1.896 = 1.138,$$
$$\text{Imperial Valley SRSS: } C_2 = 0.313 \times 1.896 = 0.593, \text{ and}$$
$$\text{Northridge SRSS: } C_3 = 0.168 \times 1.896 = 0.318.$$

The same scale factor (e.g., 1.138 for San Fernando) is applied to each component of the ground motion, and the two scaled components would be applied simultaneously in the analysis.

As with the 2D scaling, the 3D scaling results in highly amplified spectral ordinates in the range of the structure's 1.5-s fundamental period.

Comments on Ground Motion Scaling

This example has illustrated only one interpretation (the author's) of the ASCE 7 ground motion scaling requirements. Any analyst who uses the illustrated approach obtains the same set of scale factors for the same suite of ground motions. This result occurs because of the intermediate step in which the FP scale factors are applied. ASCE 7 does not require this step, and for this reason, it is possible for different analysts to obtain different ground motion scale factors when the ASCE 7 procedure is used without the FP scaling.

The 2D and 3D scaling for the structure with a period of 1.5 s produced final scale factors with highly amplified spectral ordinates at the structure's fundamental period. This result occurs because the S scaling was controlled by the portion of the spectrum in the 0.3- to 0.4-s range. The amplification factors would likely have been somewhat lower if more than three ground motions were used in the analysis because the average ground motion spectrum becomes smoother as the number of ground motions increases.

It would, of course, also be possible to obtain more favorable scale factors by choosing different ground motions. With some effort, finding three to seven ground motions for which the average spectrum is similar to the design spectrum in the period range of 0.2 to 1.5 times the fundamental period would be possible.

As a final note, ground motion selection and scaling are part of a complex process and should not be attempted without assistance from an experienced engineering seismologist.

Example 7

Selection of Structural Systems

This example examines all viable structural steel systems for two buildings in an office complex. The buildings have the same plan layout, but have different heights. One building is 10 stories tall, and the other building is four stories tall. The end of this example includes a few comments about bearing wall systems.

The best structural system for a particular building depends on many factors, such as architectural and functional requirements, labor, fabrication, and construction costs. These issues are so variable that they cannot be addressed in this guide. Another consideration in the system selection is that ASCE 7 places strict restrictions on the type and height of structural system that may be used for seismic resistant design. These restrictions are discussed in this example.

Section 12.2.1, together with Table 12.2-1, provides the basic rules for the selection of seismic force–resisting systems. Table 12.2-1 is divided into broad categories, such as bearing wall systems, building frame systems, and dual systems, and it provides system types within each category. Limitations are placed on the use of each system in terms of Seismic Design Category (SDC) and height. For example, ordinary steel concentrically braced frames (system 3 in the building frame systems category) are not permitted (NP) in SDC F, are allowed only up to heights of 35 ft in SDC D and E, and are allowed with no height limit (NL) in SDC B and C.

For each system, three design parameters are specified:
- response modification coefficient R,
- system overstrength factor Ω_o, and
- deflection amplification factor C_d.

Whereas each of these parameters affects system economy, the most influential factor is R because the design base shear is inversely proportional to this parameter via the seismic response coefficient C_s. This coefficient is specified in Section 12.8. There are basically two sets of C_s equations. The first set of equations, given by Eqs. (12.8-2), (12.8-3), and (12.8-4), represent three branches of the inelastic design response spectrum. Each equation contains R in the denominator, so it appears that larger values of R result in lower values of design base shear.

The second set of equations provides the minimum values for C_s. The first of these, Eq. (12.8-5), applies when the mapped spectral acceleration, S_1, is less than $0.6\,g$. This equation

$$C_s = 0.044 S_{DS} I_e \geq 0.01$$

is not a direct function of R but will control over Eq. (12.8-3) whenever the fundamental period of vibration is greater than $22.7\,T_S/R$.

To illustrate the use of the base shear equations, the following example uses the building shown in **Fig. G7-1**. A four-story and a 10-story version of the building are considered. The four-story building has the same floor plan and first-story height as the taller building. Based on the building use, the Risk Category of the building is II. The structure is located in an area of somewhat high seismicity and is situated on Site Class C soils. The ground motion parameters are as follows:

$S_S = 0.75\,g$,
$S_1 = 0.22\,g$,
$T_L = 6.0\,s$,
$F_a = 1.1$ (**Table GA-1** or Table 11.4-1),
$F_v = 1.58$ (**Table GA-2** or Table 11.4-2),
$S_{DS} = (2/3)(1.1)(0.75) = 0.55\,g$ [Eqs. (11.4-1) and (11.4-3)],
$S_{D1} = (2/3)(1.58)(0.22) = 0.23\,g$ [Eqs. (11.4-2) and (11.4-4)],
W (10-story building) = 22,000 kip, and
W (four-story building) = 8,800 kip.

Fig. G7-1

Base building for system comparison examples

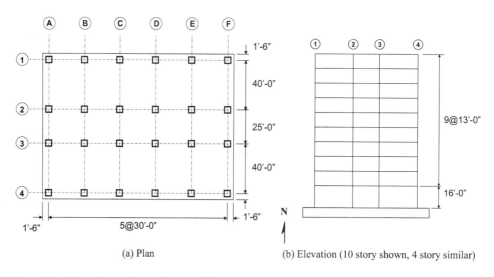

(a) Plan   (b) Elevation (10 story shown, 4 story similar)

Based on Table 1.5-2, the importance factor I_e is 1.0. In accordance with Section 11.6 and Tables 11.6-1 and 11.6-2, the building is assigned to SDC D.

A survey of locally available materials and labor costs has indicated that the structural system should be constructed from steel. Any structural system may be used, but architectural considerations require that no shear walls or diagonal bracing elements be used at the perimeter of the building.

Tables G7-1 and G7-2 list all viable structural systems for the 10-story and the four-story buildings, respectively. These systems are taken directly from Table 12.2-1 of ASCE 7. Tables G7-1 and G7-2 also list the height limit, the system R values, the period of vibration $T = C_u T_a$, the seismic response coefficient C_s, the design base shear V, and the effective R value, which is defined later in this example. The C_s values are given as computed using Eqs. (12.8-3) and (12.8-5). The period of vibration is taken as $T = C_u T_a$, with $C_u = 1.47$ for all systems. T_a is based on Eq. (12.8-7) and Table 12.8-2.

The base shear was computed using Eq. (12.8-1):

$$V = C_s W$$

Several equations are provided in the standard for computing C_s. Eq. (12.8-2) applies only if the building period is less than $T_S = S_{D1}/S_{DS} = 0.23/0.55 = 0.418\,s$, which is not the case for either of the buildings examined. Eq. (12.8-4) is not applicable because none of the building periods exceed T_L, which is 6.0 s. Eq. (12.8-6) is not applicable because S_1 is $< 0.6\,g$. This elimination leaves only Eqs. (12.8-3) and (12.8-5):

Table G7-1 Comparison of Structural Systems (10-Story Building)

System Number	Structural System	Height Limit (ft)	R	$C_u T_a$ (s)	C_s Eq. (12.8-3)	C_s Eq. (12.8-5)	V (kip)	R_{eff}
B-1	Eccentrically braced frame	160	8	1.73	0.0166	**0.0242**	532	5.5
B-2	Special concentrically braced frame	160	6	1.15	**0.0333**	0.0242	732	6.0
B-25	Buckling-restrained braced frame	160	8	1.73	0.0166	**0.0242**	532	5.5
B-26	Special steel plate shearwall	160	7	1.15	**0.0285**	0.0242	627	7.0
C-1	Special steel moment frame	No limit	8	2.06	0.0140	**0.0242**	532	4.6
C-2	Special steel truss moment frame	160	7	2.06	0.0160	**0.0242**	532	4.6
D-1	Dual system with steel EBF	No limit	8	1.15	**0.0250**	0.0242	550	8.0
D-2	Dual system with steel CBF	No limit	7	1.15	**0.0285**	0.0242	627	7.0
D-12	Dual system with BRB	No limit	8	1.15	**0.0250**	0.0242	550	8.0
D-13	Dual system with steel plate shearwall	No limit	8	1.15	**0.0250**	0.0242	550	8.0

Notes: The system number is the same as designated in Table 12.2-1 of ASCE 7. The period computed for the special steel truss moment frame is assumed to be the same as that for a standard moment frame. EBF: eccentrically braced frame; CBF: concentrically braced frame; and BRB; buckling-restrained brace. Values in bold text control the design.

Table G7-2   Comparison of Structural Systems (Four-Story Building)

System Number	Structural System	Height Limit (ft)	R	C_uT_a (s)	C_s Eq. (12.8-3)	C_s Eq. (12.8-5)	V (kip)	R_{eff}
B-1	Eccentrically braced frame	160	8	0.89	**0.0323**	0.0242	284	8
B-2	Special concentrically braced frame	160	6	0.59	**0.0646**	0.0242	568	6
B-25	Buckling-restrained braced frame	160	8	0.89	**0.0323**	0.0242	284	8
B-26	Special steel plate shearwall	160	7	0.59	**0.0553**	0.0242	487	7
C-1	Special steel moment frame	No limit	8	1.02	**0.0283**	0.0242	249	8
C-2	Special steel truss moment frame	160	7	1.02	**0.0323**	0.0242	285	7
D-1	Dual system with steel EBF	No limit	8	0.59	**0.0484**	0.0242	426	8
D-2	Dual system with steel CBF	No limit	7	0.59	**0.0553**	0.0242	487	7
D-12	Dual system with BRB	No limit	8	0.59	**0.0484**	0.0242	426	8
D-13	Dual system with steel plate shearwall	No limit	8	0.59	**0.0484**	0.0242	426	8

Notes: The system number is the same as designated in Table 12.2-1 of ASCE 7. The computed period for the special steel truss moment frames is assumed to be the same as that for a standard moment frame. EBF: eccentrically braced frame; CBF; concentrically braced frame; and BRB; buckling-restrained brace. Values in bold text control the design.

$$C_S = \frac{S_{D1}}{T\left(\dfrac{R}{I_e}\right)}$$ (Eq. 12-8-3)

$$C_S = 0.044S_{DS}I_e \geq 0.01$$ (Eq. 12.8-5)

Eq. (12.8-5) controls when $T > T_{MF}$, where

$$T_{MF} = 22.7 \, T_S/R$$

Physically, T_{MF} is the period at which Eqs. (12.8-3) and (12.8-5) give the same values for C_s.

For the 10-story building, C_s, given by Eq. (12.8-5), controls for the eccentrically braced frames, the buckling-braced frames, and the moment frames. This result occurs because of the high R values and the relatively long period of vibration for these systems. Eq. (12.8-5) does not control for the dual systems because these systems, which have high R values, have relatively low periods of vibration.

The effective R value given in column 9 of **Table G7-1** is the value of R that produces the controlling base shear. For example, for the first system listed, the controlling base shear is 532 kip. Using Eqs. (12.8-1) and (12.8-3), R_{eff} is computed as follows:

$$R_{eff} = \frac{S_{D1}I_e}{VT}W = \frac{0.23(1.0)}{532(1.73)} = 5.5$$

R_{eff} is used only in this example, and its purpose is limited to the comparisons in the example. The term R_{eff} is not used in ASCE 7.

For the systems with the base shear controlled by Eq. (12.8-5), the effective R value is less than the tabulated R value (from Table 12.2-1).

For the four-story building, all base shears are controlled by Eq. (12.8-3), so the effective R values are the same as the tabulated values.

The point to make about the analyses shown in **Tables G7-1** and **G7-2** is that for systems with high tabulated R values and relatively high periods, Eq. (12.8-5) likely controls the base shear and the apparent benefit of the high R value is lost.

The analyses in **Tables G7-1** and **G7-2** also have implications related to the calculation of drift. This issue is explored in Example 19. Additional commentary related to the minimum base shear is provided in Example 18.

Steel Frame Systems not Specifically Detailed for Seismic Resistance

Part H of Table 12.2-1 provides design values and system limitations for steel systems not specifically detailed for seismic resistance. These systems may be designed using the *Specification for Structural Steel Buildings* (AISC 2010a) and do not rely on the requirements of the *Seismic Provisions for Structural Steel Buildings* (AISC 2010b). In some cases $R = 3$ systems may be found to be more economical than systems with higher R values (e.g., ordinary steel moment frames with $R = 3.5$) that are allowed for use in the same Seismic Design Category.

Bearing Wall Systems

Section 11.2 defines bearing wall systems (under the definition for "wall") as systems in which bearing walls support all or major parts of the vertical load. Presumably, a major portion would be more than 50% of the total vertical load. Bearing walls are defined as (1) a "metal or wood stud wall that supports more than 100 lb/linear ft of vertical load in addition to its own weight," or (2) a "concrete or masonry wall that supports more than 200 lb/linear ft of vertical load in addition to its own weight." Given this definition, the likelihood is that most structural walls would be classified as bearing walls.

Consider the system shown in **Fig. G7-2**. This system, one story high, has precast concrete walls around the perimeter. The roof framing consists of steel interior tube columns and steel beams and joists. There are also steel columns between the walls on the east and west faces of the building, and these columns support the steel beams. The walls on the north and south side of the building, designated by "B," are clearly bearing walls. The walls on the east and west faces would be classified as bearing walls if the loading delivered into these walls by the roof deck were to exceed 200 lb/linear ft. For the purpose of this example, the tributary loading is assumed to be less than 200 lb/linear ft, and these walls are not designated as bearing walls.

Fig. G7-2
Concrete shear wall
system, scheme 1: not
a bearing wall system

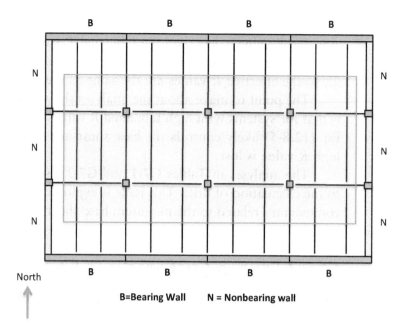

North

B=Bearing Wall   N = Nonbearing wall

The tributary vertical gravity load carried by the columns is shown by the shaded region in the interior of the structure. This region represents approximately 65% of the total load, so the bearing walls carry only 35% of the total vertical load. Hence, by definition, this system is not a bearing wall system and would be classified as a building frame system.

In **Fig. G7-3**, the system is changed such that there is only one line of columns in the interior and the center walls on the east and west side of the building support reactions from the steel girder. Hence, these walls become bearing walls. The shaded region, representing the tributary vertical load carried by the columns, is slightly more than 50% of the total area, so the system would still be classified as a building frame system.

If the steel columns were removed in their entirety and the joists spanned the full width of the building, as shown in **Fig. G7-4**, the building would be classified as a bearing wall system for loads acting in the east–west

Fig. G7-3
Concrete shear wall
system, scheme 2: not
a bearing wall system

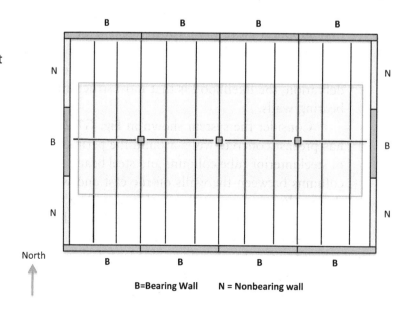

North

B=Bearing Wall   N = Nonbearing wall

Fig. G7-4

Concrete shear wall system, scheme 3: bearing wall system

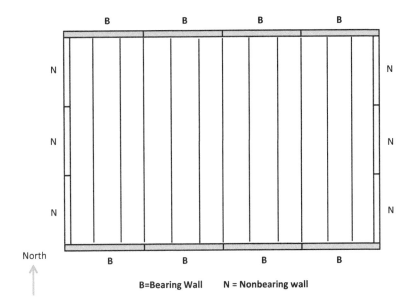

B=Bearing Wall   N = Nonbearing wall

North

direction. For loads acting in the north–south direction, the walls resisting the lateral load are not bearing walls, and the system could be classified as a nonbearing wall system for loads acting in this direction.

The main difference between bearing wall systems and building frame systems (which encompass the same lateral load-resisting elements) is that the *R* values for bearing wall systems are generally lower than those for the corresponding building frame system. Additionally, there are no bearing wall systems in structural steel, except for light-framed systems with steel sheets or light strap bracing.

Additional discussion related to bearing wall systems may be found in Ghosh and Dowty (2007).

Example 8

Combinations of Lateral Load-Resisting Systems

The examples in this chapter deal with the situation in which different lateral load–resisting systems occur in the same direction, in orthogonal directions, or along the height of a building. At issue is the height limitation for the combined system and the appropriate values of R, Ω_o, and C_d to use for the systems.

Combinations of Framing Systems in Different Directions

Where different seismic force–resisting systems are used in different (orthogonal) directions of a building and where no interaction exists between these systems, the system limitations set forth in Table 12.2-1 apply independently to the two orthogonal directions. The exception, of course, is that the lowest height limitation among all utilized systems controls for the whole building.

Consider, for example, **Fig. G8-1**, which shows plan views of simple buildings with special reinforced concrete (RC) moment frames, special RC shear walls, or a combination of systems. The lateral load resisting systems shown extend the full height of the buildings. Each building is assigned to Seismic Design Category D. The shear wall systems are considered building frame systems and not bearing wall systems.

Fig. G8-1

Plan-wise combinations
of structural systems

Special R/C Moment Frame

Special R/C Shear Wall

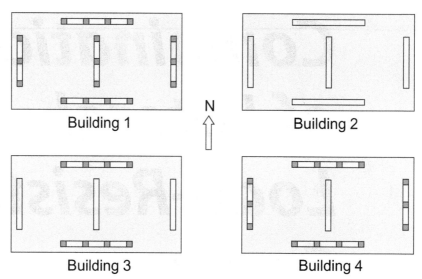

Building 1   N ↑   Building 2

Building 3   Building 4

Buildings 1 and 2 have only one lateral system, and the system limitations and design parameters are taken directly from Table 12.2-1:

Building 1: Special RC Moment Frame in Both Directions

Height limit for SDC D = None,
$R = 8$,
$\Omega_o = 3$, and
$C_d = 5.5$.

Building 2: Special RC Shear Wall in Both Directions

Height limit for SDC D = 160 ft,
$R = 6$,
$\Omega_o = 2.5$, and
$C_d = 5.0$.

The height limit for Building 2 could be increased to 240 ft if the requirements of Section 12.2.5.4 are met.

Building 3: Special RC Moment Frame in the East–West Direction and Special RC Shear Wall in the North–South Direction

Building 3 has different systems in the two different directions, but there is only one system in each individual direction. Clearly, the height limitation for the building is controlled by the shear wall. The design values for each system are as specified for that system in Table 12.2-1. The values for Building 3 are as follows:

Height limit = 160 ft (240 ft if Section 12.2.5.4 applies).

For Special RC Moment Frame in the East–West Direction

$R = 8$,
$\Omega_o = 3$, and
$C_d = 5.5$.

For Special RC Shear Wall in the North–South Direction

$R = 6$,
$\Omega_o = 2.5$, and
$C_d = 5.0$.

Building 4: Special RC Moment Frame in the East–West Direction, Combination of Special RC Moment Frame and Special RC Shear Wall in the North–South Direction, not Designed as a Dual System in the North–South Direction

Building 4 has a single system in the east–west direction and a combination of systems in the north–south direction. Assuming that the combined system is not designed as a dual system, Section 12.2.3 states that the more stringent system height limitation must be used in the north–south direction, hence the shear wall would control and the height limitation for the structure would be 160 ft.

The design parameters for the moment frames acting in the east–west direction would be taken directly from Table 12.2-1. For the north–south direction, Section 12.2.3.3 states that the value of R used for the combined system would be not greater than the least value of R for any system used in the given direction. For the structure under north–south loading, the shear wall has the lowest R, and thus $R = 6$ is assigned. Section 12.2.3.3 further stipulates that the C_d and Ω_o values for the combined system would be taken from the system that governs R, thus again, the shear wall values control. The design values for Building 4 are summarized as follows:

Height limit = 160 ft.

For the Moment Frames in the East–West Direction

$R = 8$,
$\Omega_o = 3$, and
$C_d = 5.5$.

For the Combined System in the North–South Direction

$R = 6$,
$\Omega_o = 2.5$, and
$C_d = 5.0$.

Designing the combined moment frame–shear wall as a dual system would likely be beneficial. The dual system has no height limitation, and the R value is 7, compared with that for the combined (nondual) system, which has $R = 6$. However, the moment frame in the dual system must be designed to resist at least 25% of the design base shear. The parameters for Building 4 with the north–south direction designed as a dual system are as follows:

Building 4: Special RC Moment Frame in the East–West Direction, Combination of Special RC Moment Frame and Special RC Shear Wall in the North–South Direction, Designed as a Dual System in the North–South Direction

Height limit = None.

For the Moment Frames in the East–West Direction

$R = 8$,
$\Omega_o = 3$, and
$C_d = 5.5$.

For the Dual System in the North–South Direction

$R = 7$,
$\Omega_o = 2.5$, and
$C_d = 5.5$.

Combinations of Structural Systems in the Vertical Direction

Section 12.2.3.1 provides the requirements for buildings with different systems in the vertical direction. This situation is shown for Buildings C and D in **Fig. G8-2**. Building C has an X-braced system on the bottom six levels and a moment frame on the top six levels. Building D in the same figure is just the opposite, with the moment frame at the bottom and the braced frame at the top. Buildings A and B of **Fig. G8-2** consist of a moment frame or a braced frame for the full height. In all cases, the moment frame is a special steel moment-resisting frame, and the braced frame is a special steel concentrically braced frame. The height of each of the buildings is 150 ft, and the buildings are assigned to Seismic Design Category C.

Fig. G8-2
Vertical combinations of structural systems

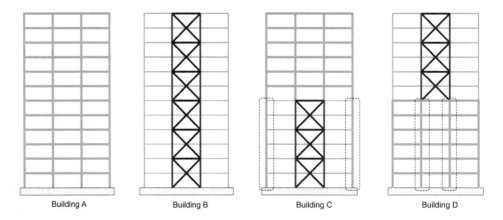

Building A   Building B   Building C   Building D

Buildings A and B have a single system along the full height, and their limitations and design parameters come directly from Table 12.2-1. The system values are summarized for these two buildings as follows.

Building A: Special Steel Moment Frame

Height limit for SDC C = None,
$R = 8$,
$\Omega_o = 3$, and
$C_d = 5.5$.

Building B: Special Steel Concentrically Braced Frame

Height limit for SDC C = None,
$R = 6$,
$\Omega_o = 2.0$, and
$C_d = 5.0$.

Building C has a special steel concentrically braced frame ($R = 6$) in the bottom six stories and a special steel moment frame ($R = 8$) in the top six stories. Section 12.2.3.1 provides the requirements for selecting the system limitations and design parameters. Subparagraph 1 of this section applies because the lower system has a lower R value than does the upper system. Hence, the upper system is designed for the R, Ω_o, and C_d values for the upper system, and the lower system is designed for the R, Ω_o, and C_d values for the lower system. Additionally, the forces transferred from the upper system to the lower system must be multiplied by the ratio of the upper system's R value to the lower system's R value. The practical aspects of this requirement are discussed later in this chapter.

Section 12.2.3.1 does not state explicitly which height limitation would apply, but this situation is irrelevant for the given example because each system has a height limit that is greater than the height of the building. In cases where a potential exists for the height limit for one component of a vertically combined system to be less than the intended height of the building it is recommended that the height for the entire building be taken as the minimum of the limits for the different components.

The design values for Building C are as follows.

Building C: Special Steel Concentrically Braced Frame for Stories 1 through 6

$R = 6$,
$\Omega_o = 2.0$, and
$C_d = 5.0$.

Forces from upper system transferred to lower system to be multiplied by $8/6 = 1.33$.

Building C: Special Steel Moment Frame for Stories 7 through 12

$R = 8$,
$\Omega_o = 3.0$, and
$C_d = 5.5$.

There could be some question as to how the exterior columns in the lower six stories of Building C would be designed. These columns are shown within the dotted line regions of **Fig. G8-2** (Building C). Strictly speaking, these columns are part of the special moment frame because they transmit the overturning moment of the upper six levels to the base of the building. These columns should be detailed as special moment frame columns.

Building D has a special moment frame in the lower six stories and special steel CBF in the upper six stories. According to Section 12.2.3.1, the entire structure should be designed using the R, Ω_o, and C_d values for the upper system. Hence, the design values for the structure are as follows:

Building D: Special Steel Moment Frame for Stories 1 through 6

$R = 6$,
$\Omega_o = 2.0$, and
$C_d = 5.0$.

Building D: Special Steel Concentrically Braced Frame for Stories 7 through 12

$R = 6$,
$\Omega_o = 2.0$, and
$C_d = 5.0$.

As with Building C, a question arises with regard to the design of the discontinuous columns in Building D. Here, the interior columns are at issue. These columns are enclosed by dotted lines in **Fig. G8-2** (Building D). Section 12.3.3.3 requires that elements supporting discontinuous frames be designed using load factors that include the overstrength factor Ω_o when a Type 4 vertical irregularity (in-plane discontinuity irregularity) exists. It is not clear whether an irregularity exists because no apparent offset occurs in the lateral load–resisting system and because it is unclear whether a reduction in stiffness occurs below the transition.

The author would not be inclined to use the Ω_o factors for the exterior columns of Building C, but would use the Ω_o factors for the interior columns of Building D.

As a final point, ASCE 7 places severe restrictions on systems similar to Building C in **Fig. G8-2** (moment frames above braced frames) when such systems are used in buildings assigned to SDC D and higher. These restrictions, given in Section 12.2.5.5, are not applicable to the buildings in this example because the SDC was C.

Approximate Periods of Vibration for Combined Systems

Determining the approximate period of vibration, T_a, for a structure is almost always necessary. This period is used in computing the seismic base shear when the equivalent lateral force (ELF) method of analysis is used and for scaling the results of a modal response spectrum analysis when the base shear from such an analysis is less than 85% of the ELF base shear (Section 12.9.4). The approximate period is computed using Eq. (12.8-7), which uses the parameters C_t and x. Table 12.8-2 provides these parameters for several well defined systems but does not specifically address combined or dual systems. Hence, dual systems or systems combined in the same direction would apparently fall under the "All other structural systems" category. This approach seems overly conservative, and using a weighted average based on the parameters in Table 12.8-2 seems reasonable. For example, the period for Building D of **Fig. G8-2** could be estimated as the average of the period of a 12-story moment frame and a 12-story braced frame. However, the average of the periods of a six-story moment frame (the top half) and a six-story braced frame (the bottom half) would not be appropriate. Nor would it be appropriate to use one period for the upper half of the building and a different period for the lower half.

Structural Analysis for Combined Systems

When designing buildings with mixed R values, it is recommended that the analysis be performed with $R = 1$ and that the actual R values be assigned on a member-by-member basis. In this case, interstory drifts would be multiplied by C_d/R, where again, the actual R value would be used.

Vertical Combination when the Lower Section Is Stiff Relative to the Upper Portion

Where the lower portion of a building is much stiffer than the upper portion, a two-stage ELF procedure is permitted to determine design forces. Section 12.2.3.2 stipulates that the two-stage analysis is limited to systems in which the lower portion is at least 10 times as stiff as the upper portion and for which the period of the entire structure is not greater than 1.1 times the period of the upper portion of the structure, with the upper portion fixed at its base. A typical building that exhibits this configuration is a multistory wood frame apartment over a one-story concrete shear wall parking structure.

The stiffness requirements are difficult to apply because the standard does not specify how the stiffness is to be computed and several measures of the stiffness of a structure exist. Because of these complexities, examples of this type of system are deferred to Example 18, which covers the equivalent lateral force method.

Example 9

Horizontal Structural Irregularities

Section 12.3.2.1 is used to determine if one or more horizontal structural irregularities exist in the lateral load–resisting system. The five basic irregularity types are described in Table 12.3-1. This example explores these irregularities, but concentrates primarily on horizontal structural irregularity Types 1a and 1b (torsion and extreme torsion).

Torsional Irregularities (Types 1a and 1b)

The definitions in Table 12.3-1 indicate that a torsional irregularity exists if the story drift, Δ, at one end of a building is more than 1.2 times the average of the drift at the two ends of the building when that building is subjected to lateral forces applied at the accidental torsion eccentricity of 0.05 times the length of the building. If the ratio of story drifts is greater than 1.4, an extreme torsional irregularity exists. Torsional irregularity must only be checked for systems with rigid or semirigid diaphragms. If a torsional irregularity exists at any level of such a building for loads acting in any direction, the entire structure is torsionally irregular.

Two examples are provided for determining if a torsional irregularity exists. In the first example, a simple one-story structure with symmetrically placed shear walls is analyzed to determine the effect of the placement of the walls on the torsional stiffness of the system. The second example is more realistic in the sense that it determines whether torsional irregularities occur in a typical four-story office building.

Consider first the simple one-story system shown in **Fig. G9-1**. The lateral system for the building consists of four walls, each with an in-plane lateral stiffness k. The out-of-plane lateral stiffness is assumed to be zero.

Fig. G9-1

Simple building for
evaluation of torsional
irregularity

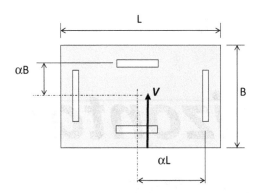

The walls are placed symmetrically in the system; the parameter α is used to locate the walls some distance from the center of mass, which is located at the geometric center of the building. The diaphragm is assumed to be rigid. The lateral load V is applied at an accidental eccentricity of 0.05 times the building width, L.

Given this configuration, the displacement at the center of the building is

$$\Delta_{\text{CENTER}} = \frac{V}{2k} \qquad \text{(Eq. G9-1)}$$

and the deflection at the edge of the building is

$$\Delta_{\text{EDGE}} = \frac{V}{2k} + \frac{0.05VL}{2k(\alpha^2 L^2 + \alpha^2 B^2)}(0.5L) \qquad \text{(Eq. G9-2)}$$

The deflection at the center of the building is the same as the average of the deflections at the extreme edges of this rigid diaphragm building.

Using Eqs. (G9-1) and (G9-2) and simplifying, the ratio of the displacement at the edge to the center is

$$\text{Ratio} = \frac{\Delta_{\text{EDGE}}}{\Delta_{\text{CENTER}}} = 1 + \frac{0.025}{\alpha^2 \left[1 + \left(\dfrac{B}{L}\right)^2\right]} \qquad \text{(Eq. G9-3)}$$

Fig. G9-2 plots Eq. (G9-3) for four values of B/L and for α values ranging from 0.1 (all walls near the center of the building) to 0.5 (all walls on the perimeter of the building). Also shown are the limits for a torsional irregularity (ratio = 1.2) and for an extreme torsional irregularity (ratio = 1.4). For the rectangular building with $B/L = 0.25$, the torsional irregularity occurs when α is approximately 0.35 and the extreme irregularity occurs when $\alpha = 0.24$. For the square building ($B/L = 1$), the torsional and extreme torsional irregularities occur at $\alpha = 0.25$ and 0.17, respectively.

Three important observations may be drawn from the results:

1. Torsional irregularities may occur even when the lateral load–resisting system is completely symmetric (no inherent torsion).
2. The closer the walls are to the center of the building (lower values of α), the greater the possibility of encountering a torsional

Fig. G9-2
Effect of wall
placement on
torsional irregularity

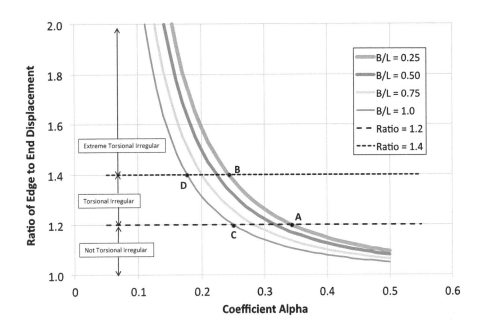

irregularity. Thus, buildings with interior reinforced concrete core-walls or interior core bracing as the only lateral load–resisting system are very likely to be torsionally irregular.

3. Torsional irregularities are more likely to occur in rectangular buildings than in square buildings.

Section 12.8.4.3 of ASCE 7 requires that the accidental torsion be amplified in Seismic Design Categories C, D, E, and F. Note that this amplification is required only where accidental torsion is applied as a static loading. Where dynamic analysis is used and accidental torsion is accounted for by using a physical mass eccentricity, amplification of accidental torsion is not required. See Section 12.9.5. The amplification factor A_x is given as follows:

$$A_x = \left(\frac{\delta_{max}}{1.2\delta_{avg}}\right)^2 \leq 3.0 \qquad \text{(Eq. 12.8-14)}$$

Fig. G9-3 plots the amplification factors for the same four B/L ratios discussed previously. Also shown on the plot is a line with a constant value of 3.0, which represents the maximum amplification factor that needs to be used in analysis. For the rectangular building with $B/L = 0.25$, the maximum amplification occurs when α is 0.15, and for the square building, the maximum occurs when α is 0.11.

The second example for evaluating torsional irregularity is based on a six-story building with a typical floor plan as shown in **Fig. G9-4**. The lateral load–resisting systems consist of moment frames on grid lines A and D and braced frames on lines 2, 3, 4, and 5. There is no bracing on grid line 6, so the center of mass and the center of rigidity do not coincide with respect to loading in the transverse direction. The height of the first story is 16 ft, and the height of stories 2 through 6 is 13 ft, giving a total building height of 81 ft.

The results of the torsional analysis for loading in the transverse direction are shown in **Table G9-1**. The analysis was run assuming a rigid

Fig. G9-3

Effect of wall placement on torsional amplification

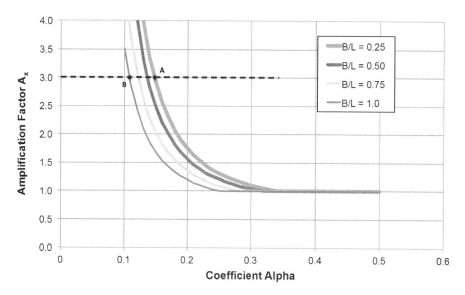

Fig. G9-4

Plan view of six-story building for evaluation of torsional irregularity

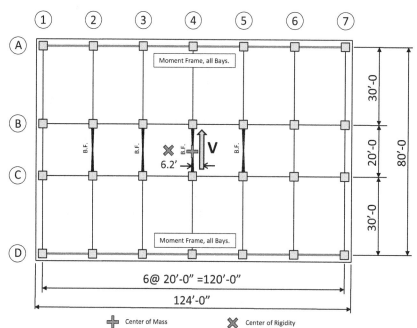

Table G9-1 Torsional Irregularity Check for the Building in **Fig. G9-4**

Level	F_i (kip)	T_{ai} (kip-ft)	δ_i Line 4 (in.)	Δ_i Line 4 (in.)	δ_i Line 7 (in.)	Δ_i Line 7 (in.)	Ratio (7/4)	Calculated A_x
6	207.0	1,283.3	3.219	0.601	3.863	0.685	1.139	1.000
5	160.1	992.6	2.618	0.605	3.178	0.707	1.168	1.023**
4	111.9	693.5	2.013	0.501	2.471	0.607	1.212*	1.047
3	70.9	439.7	1.512	0.472	1.864	0.583	1.235*	1.056
2	37.9	235.0	1.040	0.464	1.281	0.588	1.268*	1.054
1	15.0	93.0	0.576	0.576	0.693	0.693	1.203*	1.005

*Result is greater than 1.2 so a Type 1a torsional irregularity exists.

**An amplification factor of 1.0 could be applied at this level because no torsional irregularity exists.

diaphragm. Column 2 of the table shows the equivalent lateral forces at each level computed according to the equivalent lateral force procedure of Section 12.8. Column 3 of the table provides the accidental torsions at each level, which are equal to the story force times 6.2 ft (0.05 times the width of the building). The torsions are based on the lateral forces applied to the right of frame line 4 because this produces the largest displacements at the edge of the building. (In a more complicated building, this is not apparent, and both directions should be checked for eccentricity.) Column 4 of the table provides the computed story displacements at grid line 4, and column 5 lists the interstory drifts on grid line 4. Columns 6 and 7 contain the displacements and interstory drifts at the edge of the building (2.0 ft to the right of grid line 7). The ratio of the drift at the edge of the building to the drift at the center is shown in column 8. Column 8 shows that the ratio of edge drift to center drift exceeds 1.2, but is less than 1.4, for stories 1 through 4, so the structure has a Type 1a torsional irregularity. This building is very close to being torsionally regular, and thus decreasing the stiffness in frame 2, or adding some bracing to frame 6 would be advisable as this would eliminate the irregularity.

The torsional amplification factors, computed according to Eq. (12.8-14), are listed in column 9 of **Table G9-1**. These quantities are based on story displacements, not interstory drifts. For level 3, for example,

$$A_x = \left(\frac{\delta_{max}}{1.2\delta_{avg}} \right)^2 = \left[\frac{1.864}{1.2(1.512)} \right]^2 = 1.056$$

Example 14 of this guide contains additional discussion of accidental torsion and torsional amplification.

Reentrant Corner Irregularity (Type 2)

According to Table 12.3-1, a reentrant corner irregularity occurs when both plan projections are greater than 15% of the width of the plan dimension in the direction of the projection. **Fig. G9-5** shows the plan of a building with four reentrant corners, marked A through D. For this building, only corner D would cause a reentrant corner irregularity because both projections are greater than 15% of the width of the building.

In some cases, a notch in the edge of the building may trigger a reentrant corner irregularity. See the following discussion on diaphragm irregularities for more details.

Diaphragm Discontinuity Irregularity (Type 3)

According to Table 12.3-1, diaphragm discontinuity irregularities occur if the area of a cutout or hole in the diaphragm is greater than 50% of the

Fig. G9-5

Building with four reentrant corners and a reentrant irregularity

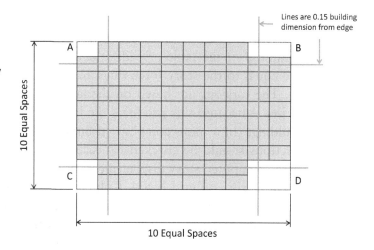

gross enclosed area of the diaphragm, or if the in-plane stiffness of the diaphragm at one level is less than 50% of the stiffness at an adjacent level.

Fig. G9-6 shows four different diaphragms. In **Fig. G9-6(a)**, the opening is a notch that is less than 50% of the gross enclosed area, so the diaphragm is not irregular. Note the presence of the exterior window wall, which is required if the shown opening is to be considered as part of the enclosed area. If this window wall did not exist, the opening might in fact cause a reentrant corner irregularity because the projections caused by the opening are greater than 15% of the building width. This situation is shown in **Fig. G9-7**.

In **Figs. G9-6(b)** and **G9-6(c)**, the diaphragm openings are in the interior of the building, and neither triggers a diaphragm irregularity because the areas of the openings are less than 50% of the gross enclosed area. However, the opening in **Fig. G9-6(d)** does cause a diaphragm irregularity.

Fig. G9-6

Diaphragm openings and irregularities

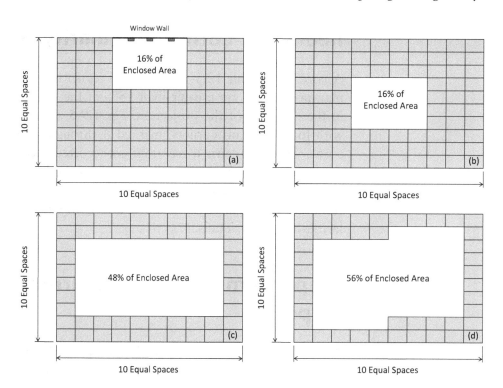

Fig. G9-7

Diaphragm notch causing a reentrant corner irregularity

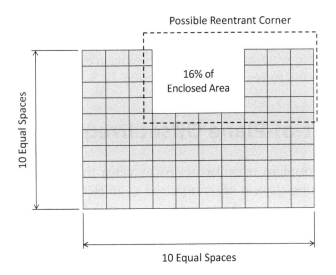

A diaphragm irregularity can also occur if the in-plane stiffness at one level is less than 50% of the stiffness at an adjacent level. Calculations to determine diaphragm stiffness are not straightforward but in some cases can be accomplished using finite element analysis. For example, a finite element model for the diaphragm in **Fig. G9-6(b)** is shown in **Fig. G9-8**. The diaphragm stiffness would be computed as the quantity V/Δ. In the figure, the lines to the left and right edges represent lines of support provided by lateral load–resisting elements. For more complex systems, with multiple diaphragm segments and multiple lateral load–resisting elements, the determination of diaphragm stiffness is essentially impossible in terms of the definition provided in Table 12.3-1.

A diaphragm irregularity based on a differing stiffness of adjacent stories may in fact be irrelevant. Consider, for example, a rectangular building with no diaphragm openings. At one level, the floor slab is 4 in. thick, and at an adjacent level, the thickness is 10 in. Clearly the 10-in.-thick diaphragm is more than twice as stiff as the 4-in.-thick diaphragm, but in terms of stiffness relative to the lateral load–resisting system, both may be considered rigid. Hence, the different stiffnesses of the diaphragms have virtually no effect on the analysis, with the exception that the increased thickness

Fig. G9-8

Finite element model for determining diaphragm stiffness

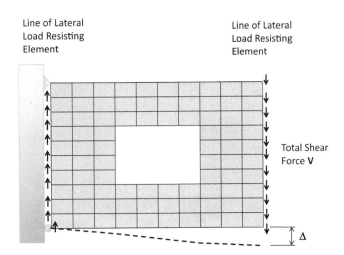

may cause a mass irregularity. However, regardless of the actual diaphragm behavior, the classification of the diaphragm as discontinuous cannot be ignored when addressing the consequences of the irregularity (e.g., increased collector forces required by Section 12.3.3.4).

Out-of-Plane Offset Irregularity (Type 4)

Out-of-plane irregularities occur when the lateral forces in a lateral load–resisting element are transferred to an element that is not in the same plane as that element. An example is shown in **Fig. G9-9**, which is a plan view of the Imperial County Services Building, which was severely damaged during the October 15, 1979, Imperial Valley earthquake. In this building, wall A, an exterior wall, occurs on stories 2 through 6 and transfers shear to wall B, which exists only on the first story. Overturning moment is transferred to the columns adjacent to but offset from wall A. The walls labeled C extend the full height of the building. The lateral system in the long direction consists of moment-resisting frames. During the earthquake, the columns adjacent to wall A but not on the same line as wall A failed because of the combined effect of overturning in the transverse direction and moment frame action in the longitudinal direction. A photograph of the building is shown in **Fig. G9-10**.

Fig. G9-9
Plan view of building with an out-of-plane offset irregularity

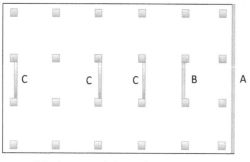

A) 2nd story and above shear wall
B) 1st story only shear wall
C) Full height shear wall

Fig. G9-10
Photograph of the Imperial County Services Building

Source: Courtesy of V. Bertero.

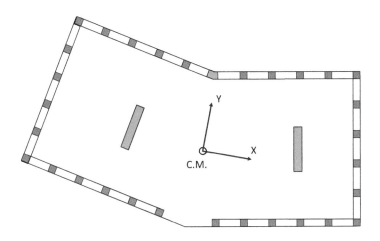

Fig. G9-11
Structure with a nonparallel system irregularity

Nonparallel System Irregularity (Type 5)

Nonparallel system irregularities occur when any element of the lateral load–resisting system is not parallel to one of the orthogonal axes of the lateral load–resisting system of the entire structure. Such a system is shown in **Fig. G9-11**, which is a plan view of a reinforced concrete frame-wall system. The axes marked *X* and *Y* represent the principal axes of the entire structural system. Clearly, the line of action of the lateral load–resisting elements is not parallel to either the *X* or the *Y* axis, thus a nonparallel system irregularity exists.

Consequences of Horizontal Irregularities

Horizontal irregularities are significant primarily when the structure under consideration is assigned to SDC D or above, or in some cases, SDC C. The third and fourth columns of Table 12.3-1 provide the consequences of the irregularities in terms of the seismic SDC. For example, a Type 5 horizontal irregularity triggers the requirement for consideration of orthogonal load effects (Section 12.5.3) in buildings assigned to SDC C and above and requires three-dimensional analysis in all SDC levels (Section 12.7.3). Additionally, Section 12.3.3.1 provides circumstances in which certain horizontal irregularities are prohibited. Finally, Table 12.6-1 (Permitted Analytical Procedures) contains restrictions on the use of the ELF analysis method when certain irregularities exist.

Example 10

Vertical Structural Irregularities

Section 12.3.2.2 is used to determine if one or more vertical structural irregularities exist in the lateral load–resisting system. The five basic irregularity types are described in Table 12.3-2. This example explores each irregularity, but concentrates primarily on irregularity Types 1a and 1b (soft story) and Types 5a and 5b (weak story).

Soft Story (Stiffness) Irregularities (Types 1a and 1b)

The first step in a soft story irregularity check is to use the first exception in Section 12.3.2.2 to determine if potential exists for a soft story irregularity. This check is based on interstory drift and is not difficult to perform. If the relative drift criteria are met, no soft story irregularity exists and the check is complete. If the drift criteria are not met, the irregularity must be accepted, or the stiffness-based check of Table 12.3-2 must be performed. The stiffness check has three possible results: no irregularity exists, a soft story irregularity exists (Type 1a), or an extreme soft story irregularity exists (Type 1b).

For the drift-based check, the structure is subjected to the design lateral loads, and interstory drift ratios are computed for each story. If the drift ratio in each story is less than 1.3 times the drift ratio in the story directly above it, no stiffness irregularity exists. When performing the drift check, the top two stories of the building need not be evaluated and accidental torsion need not be included.

Although ASCE 7 requires that the design-level lateral loads be used in the check, this is not strictly necessary when performing linear analysis. It

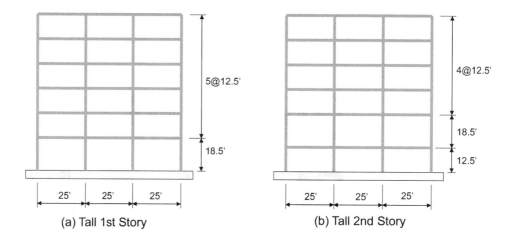

Fig. G10-1
Building used to investigate soft story irregularity

(a) Tall 1st Story   (b) Tall 2nd Story

is important, though, that the vertical distribution of the lateral loads be reasonably correct. On the basis of this concept, the lateral loads used for the stiffness irregularity check may be based on the equivalent lateral force (ELF) method described in Section 12.8. The result of the calculation may show, however, that the ELF method may not be used for the final design of the structure. Applicability of the ELF method is covered in Section 12.6 and Table 12.6-1.

This example considers a six-story reinforced concrete moment frame located in a region of high seismicity ($S_{D1} > 0.4\,g$). For simplicity, assumptions include that the story weight for each level is 1,500 kip and that all story heights are equal to 12.5 ft, except for one tall story, which has a height of 18.5 ft. Two separate analyses are run, one for which the tall story is the first story of the building, and one for which the tall story is the second story of the building (**Fig. G10-1**).

The period of vibration of the structure is estimated as

$$T = C_u T_a \qquad \text{(from Section 12.8.2)}$$

where C_u is taken from Table 12.8-1 and where

$$T_a = C_t h_n^{\ x} \qquad \text{(Eq. 12.8-7)}$$

For our structure, h_n is 81 ft, and for a concrete moment frame, $C_t = 0.016$ and $x = 0.9$ (Table 12.8-2). Hence,

$$T_a = C_t h_n^{\ x} = 0.016 \times 81^{0.9} = 0.835\,\text{s}$$

Note that $T = C_u T_a$ may only be used if a properly substantiated (computer) analysis has been used to determine the true analytical period (called T_{computed} in this *Guide*). This example assumes such an analysis has been performed and that T_{computed} exceeds $C_u T_a$, thereby setting $C_u T_a$ as the upper limit on period. This result produces a more realistic value of the exponent k than the use of T_a alone. Because of the high seismicity, $C_u = 1.4$ and

$$T = C_u T_a = 1.4 \times 0.835 = 1.17\,\text{s}$$

Based on the text in Section 12.8.3, the exponent k can be computed as

$k = 1$ when T is less than or equal to $0.5\,\text{s}$,
$k = 0.5\,T + 0.75$ when $0.5 < T < 2.5\,\text{s}$, and
$k = 2$ when T is greater than or equal to $2.5\,\text{s}$.

For the current example

$k = 0.5\,T + 0.75 = 0.5(1.17) + 0.75 = 1.33$

For computing the lateral force, a total base shear of 100 kip is assumed, and this shear is distributed vertically according to Eq. (12.8-12). **Tables G10-1** and **G10-2** show the lateral load computations for the structures with the tall first and second story, respectively.

The resulting story displacements, story drifts, story drift ratios, and ratio of story drift ratios are shown in **Tables G10-3** and **G10-4**. These computations do not include the deflection amplification factor C_d because this factor cancels out when calculating the ratios of the story drift ratios.

In **Table G10-3**, which is for the structure with the tall first story, the interstory drift ratio (IDR) in the tall bottom story (0.43%) is actually smaller than the drift ratio at the next story above (0.53%). The first-story drift ratio, divided by the second-story drift ratio is 0.811. This rather unexpected result occurs because the fixed-base condition stiffens the bottom

Table G10-1   Development of Lateral Loads for Structure with Tall First Story

Story	H (ft)	h (ft)	w (kip)	wh^k	wh^k/Total	F (kip)
6	12.5	81.0	1,500	521,128	0.306	30.6
5	12.5	68.5	1,500	416,898	0.245	24.5
4	12.5	56.0	1,500	318,811	0.187	18.7
3	12.5	43.5	1,500	227,765	0.134	13.4
2	12.5	31.0	1,500	145,080	0.085	8.5
1	18.5	18.5	1,500	72,968	0.043	4.3
			Totals	1,702,650	1.000	$V = 100.0$

Table G10-2   Development of Lateral Loads for Structure with Tall Second Story

Story	H (ft)	h (ft)	w (kip)	wh^k	wh^k/Total	F (kip)
6	12.5	81.0	1,500	521,128	0.311	31.1
5	12.5	68.5	1,500	416,898	0.249	24.9
4	12.5	56.0	1,500	318,811	0.191	19.1
3	12.5	43.5	1,500	227,765	0.136	13.6
2	18.5	31.0	1,500	145,080	0.087	8.7
1	12.5	12.5	1,500	43,297	0.026	2.6
			Totals	1,672,979	1.000	$V = 100.0$

Table G10-3   Drift-Based Soft Story Check for Structure with Tall First Story

Story	H (in.)	δ (in.)	Δ (in.)	IDR (%)	IDR_n/IDR_{n+1}
6	150	4.29	0.41	0.27	—
5	150	3.88	0.63	0.42	1.56
4	150	3.25	0.75	0.50	1.19
3	150	2.50	0.74	0.50	1.00
2	150	1.75	0.79	0.53	1.06
1	222	0.96	0.96	0.43	0.811

Note: $IDR = \Delta/H \times 100\%$.

Table G10-4   Drift-Based Soft Story Check for Structure with Tall Second Story

Story	H (in.)	δ (in.)	Δ (in.)	IDR (%)	IDR_n/IDR_{n+1}
6	150	4.60	0.42	0.28	—
5	150	4.18	0.65	0.43	1.54
4	150	3.54	0.78	0.52	1.21
3	150	2.75	0.82	0.55	0.96
2	222	1.93	1.47	0.66	1.20
1	150	0.46	0.46	0.31	0.47

story relative to the upper stories. This effect may be seen in the deflected shape profile, which is presented in **Fig. G10-2(a)**.

The maximum ratio of IDRs in **Table G10-3** is 1.56, which is for the fifth story relative to the sixth story. Although this ratio is greater than 1.3, it does not result in a soft story classification because the first exception in Section 12.3.2.2 states that the top two stories of the structure may be excluded from the check. By the nature of this exception, though, only buildings taller than two stories are subject to this irregularity.

The drift calculations are shown for the structure with the tall second story in **Table G10-4**. Here, the second-story drift ratio (0.66%) is 1.2 times the drift ratio of the third story (0.55%). A soft story condition does not occur because this value is less than 1.3. The deflected shape profile for the structure with the soft second story is shown in **Fig. G10-2(b)**.

Based on these results, both structures in **Fig. G10-1** are exempt from the stiffness-based soft story check described in the first two rows of Table 12.3-2. However, this check may be required in some circumstances. To illustrate this procedure, a stiffness-based soft story check is performed for the building with the tall second story [**Fig. G10-1(b)**], even though such a check is not actually required for this structure.

The first step in the analysis is determining the story stiffness. This is done on a story-by-story basis by applying equal and opposite lateral forces V at the top and bottom of the story, computing the interstory drift Δ in the story, and defining the interstory stiffness as

Fig. G10-2
Deflected shape profiles

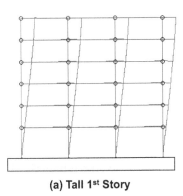

(a) Tall 1st Story

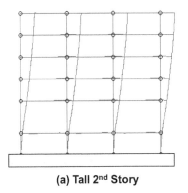

(a) Tall 2nd Story

Table G10-5

Stiffness-Based Soft Story Analysis for Structure with Soft Second Story

Story	Δ (in.)	K (kip/in.)	K_n/K_{n+1}	K_n/Avg K_{n+1}
6	0.763	131	—	—
5	0.691	145	145/131 = 1.11	—
4	0.622	161	161/145 = 1.11	—
3	0.511	196	196/161 = 1.22	196/146 = 1.35
2	1.042	96	96/196 = 0.490	96/167 = 0.575
1	0.290	345	345/96 = 3.59	345/151 = 2.28

$$K_i = \frac{V}{\Delta_i}$$

After the story stiffnesses are determined, they are compared according to the requirements of Table 12.3-2. A soft story irregularity exists if for any story the stiffness of that story is less than 70% of the stiffness of the story above, or if the stiffness of the story is less than 80% of the average stiffnesses of the three stories above. The irregularities are considered extreme if for any story the stiffness of that story is less than 60% of the stiffness of the story above, or if the stiffness of the story is less than 70% of the average stiffnesses of the three stories above.

The results for the frame of **Fig. G10-1(b)** are shown in **Table G10-5**. A story shear of $V = 100$ kip was used in the analysis.

Based on this check, the structure has an extreme soft story irregularity. However, the structure need not be classified as such because the drift-based check of the same structure exempted this structure from the stiffness-based check. Based on this observation, one should always perform the drift-based check first because this step may exempt a structure from being classified as having a soft story irregularity, thereby allowing the designer to skip the more time-consuming stiffness-based check.

Weight (Mass) Irregularity (Type 2)

Vertical weight irregularities are relatively straightforward, and no example is presented. Note, however, that the story weight used in the calculation is the effective seismic weight, as defined in Section 12.7.2. Also, the same exception (the drift ratio test) that applies to stiffness irregularities may be applied to weight irregularities. This exception may supersede the mass ratio test provided in Table 12.3-2.

Vertical Geometric Irregularity (Type 3)

A vertical geometric irregularity occurs when the horizontal dimension of the lateral-resisting system at one level is more than 130% of that for an adjacent story. Based on this definition, the structure shown in **Fig. G10-3(a)** has a geometric irregularity because the moment-resisting frame has three bays on the third story and only two bays on the fourth story. However, the structure in **Fig. G10-3(b)** does not have a vertical geometric irregularity because the braced frame, which is the lateral-resisting system, has the same horizontal dimension for the full height. The setbacks on the upper three stories have no influence on the vertical geometric irregularity of the braced frame system because the two exterior bays resist gravity loads only and, as such, are not part of the lateral load–resisting system. The moment frame may also have a stiffness irregularity, and both systems are likely to have a weight irregularity.

In-Plane Discontinuity in Vertical Lateral Force–Resisting Element Irregularity (Type 4)

An in-plane discontinuity occurs when a horizontal offset occurs in the lateral load–resisting system that causes an overturning moment demand on a supporting beam, column, truss, or slab. Based on this definition, the system in **Fig. G10-4(a)** has an irregularity because the offset produces over-turning moment demands on the columns that support the upper three stories of the X-braced frame. For the system in **Fig. G10-4(b)**, the offset again produces overturning moment demands on the two columns support-ing the upper three stories.

Fig. G10-3
Example of vertical geometric irregularity

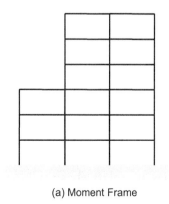
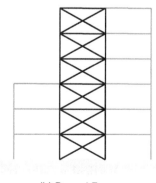

(a) Moment Frame   (b) Braced Frame

Fig. G10-4

Example of an in-plane discontinuity irregularity

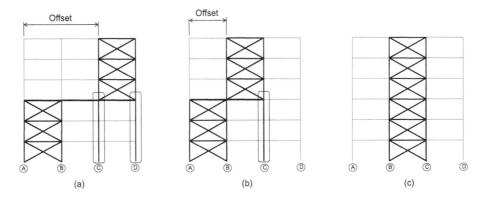

In the system in **Fig. G10-4(a)**, columns C and D (shown in the figure surrounded by a dotted line) under the discontinuous braced frame would be subject to the requirement of Section 12.3.3.3, thereby requiring them to be designed with load cases that include the overstrength factor Ω_o. Column C in **Fig. G10-4(b)** would also be subject to the requirements of Section 12.3.3.3 because the offset transfers an overturning moment to this column. However, column B in the same structure would not be required to be designed for load cases including Ω_o because this column would carry overturning forces even if the offset did not exist [**Fig. G10-4(c)**].

Discontinuity in Lateral Strength–Weak Story Irregularity (Types 5a and 5b)

Weak story irregularities are difficult to detect because the concept of story strength is not well defined, even for relatively simple systems, such as moment frames and X-braced frames. Providing a full numerical example of this type of irregularity is beyond the scope of this *Guide* because the computation of story strength depends on rules established in the material specifications, such as ACI 318 (ACI 2008) or the AISC/ANSI 341-5 (AISC 2005).

Some discussion is warranted, however. Consider the case of a moment-resisting frame shown in **Fig. G10-5**. In **Fig. G10-5(a)**, a column mechanism is assumed to have formed. This kind of mechanism can form if the columns are weak relative to the beams, which is not allowed for special moment frames in steel or reinforced concrete. The mechanism in **Fig. G10-5(b)** is a beam mechanism. As shown, this mechanism is in fact incorrect because a single-story beam mechanism cannot occur because plastic hinges would need to form in all beams at all stories. Additionally, hinges would have to form at the base of the columns (if the columns are fixed at the base). The beam mechanism could occur at one level only, however, if the columns in the story above and the story below the beam with the plastic hinges had a moment-free hinge at mid height. On the basis of these assumptions, the commentary to AISC (2005) provides formulas for computing the story strength for the mechanisms shown in **Fig. G10-5**. The intended use of the AISC expressions is related to frame stability and not system irregularity.

Fig. G10-5

Moment frame mechanisms

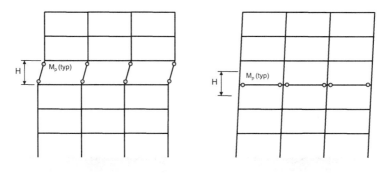

(a) Moment Frame (Story Mechanism)   (b) Moment Frame (Beam Mechanism)

For the story mechanism [**Fig. G10-5(a)**], which is applicable for systems that do not satisfy the strong column–weak beam criterion, the story strength of story, V_{yi}, is

$$V_{yi} = \frac{2\sum_{k=1}^{m} M_{pCk}}{H}$$

(Eq. C3-3, AISC 2005)

For the beam (girder) mechanism that does satisfy the strong column–weak beam requirements [**Fig. G10-5(b)**], the AISC expression for computing story strength is

$$V_{yi} = \frac{2\sum_{j=1}^{n} M_{pGj}}{H}$$

(Eq. C3-2, AISC 2005)

where k and j are integer counters,

m is the number of columns,

M_{pCk} is the plastic moment strength of column k under minimum factored load,

n is the number of bays,

M_{pGj} is the plastic moment strength of beam j, and

H is the story height.

Eqs. (C3-3) and (C3-2) (AISC, 2005) would be equally applicable to structures of reinforced concrete and are suitable for computing story strength in association with the requirements of Table 12.3-2. For steel or concrete, the effect of the axial force in the columns on moment strength of the plastic hinges must be considered.

For braced frames, the story strength depends primarily on the bracing configuration, the axial strength of the brace, and the angle of attack θ of the brace, as shown in **Fig. G10-6**. In part a, the system is a buckling-restrained braced frame, for which the strength of the single brace is the same in tension as it is in compression. For the concentrically braced frame, the strength of the tension and compression brace is different, and this difference has to be taken into consideration.

Finding the story strength of other systems, such as frame-wall systems in concrete or moment frames in combination with braced frames in steel,

Fig. G10-6
Braced frame strength irregularities

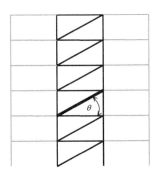 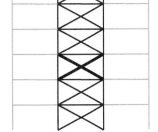

(a) Buckling Restrained Braced Frame   (b) Concentrically Braced Frame

Fig. G10-7
Computing stiffness for story *i*

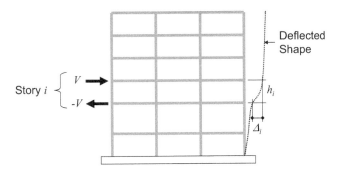

is not straightforward and must be computed using nonlinear static analysis. A loading similar to that used to determine story stiffness (**Fig. G10-7**) could be used. Such analysis should include gravity load effects if gravity loads are expected to have an influence on member strength. A rigorous analysis would include P-delta effects as well.

Weak structure irregularities are highly undesirable and should be avoided if at all possible. Fortunately, such irregularities are uncommon in structures designed according to ASCE 7 and the material specifications (e.g., AISC 2005). The rarity of the irregularity is due to the fact that the design story shears always increase from the top to the bottom of the structure, and hence the story strengths increase from the top to the bottom as well.

Consequences of Vertical Irregularities

Vertical irregularities are significant primarily when the structure under consideration is assigned to SDC D or above, or in some cases, SDC C. The second and third columns of Table 12.3-2 provide the consequences of the irregularities in terms of the seismic SDC. Section 12.3.3.1 prohibits weak story irregularities in SDC E or higher and prohibits extreme weak story irregularities in SDC D or higher.

Example 11

Diaphragm Flexibility

Roof and floor diaphragms may be classified as flexible, semirigid, or rigid. Section 12.3.1.1 describes the conditions under which a diaphragm may be considered flexible, and Section 12.3.1.2 establishes the conditions under which the diaphragm may be considered rigid. If the diaphragm cannot be classified as flexible or rigid under these rules, an analytical procedure described in Section 12.3.1.3 is required. This procedure results in the classification of the diaphragm as either flexible or semirigid. This example demonstrates the analytical procedure and illustrates the methodology for determining if a torsional irregularity exists for a system with a semirigid diaphragm. The descriptions under which a diaphragm is flexible according to Section 12.3.1.1 are not always realistic, and, indeed, a flexible diaphragm does not really exist. Most diaphragms, even untopped metal deck or wood diaphragms with length-to-depth ratios up to 4.0, tend to be more rigid than flexible. In this context, ASCE 7 never requires a diaphragm to be modeled as flexible; however, in some cases it does permit it.

The structure to be considered in the example is shown in plan and elevation in **Figs. G11-1(a)** and **G11-1(b)**, respectively. The purpose of the structure is storage for hazardous chemicals. The lateral load–resisting system in the transverse directions consists of four reinforced concrete shear walls, each 10 in. thick. The diaphragm, also constructed from concrete, is 4 in. thick. The lateral load–resisting system in the longitudinal direction consists of 10-in. walls, 11 ft long, placed at the center of each bay.

The length of the diaphragm between walls is 44 ft, and the depth of the diaphragm is 12 ft, producing a span-to-depth ratio of 3.67. According to Section 12.3.1.2, the diaphragm cannot automatically be considered rigid because the span-to-depth ratio is greater than 0.3[1]. An analysis must be

[1] Figure 12.3.1 of ASCE 7 indicates that the span of a diaphragm should be taken as the distance between lateral load–resisting elements. This rather unrealistic example was devised to produce a span-to-depth ratio greater than 3.0 for a single diaphragm segment.

Fig. G11-1
Three-bay concrete
structure analyzed for
diaphragm flexibility

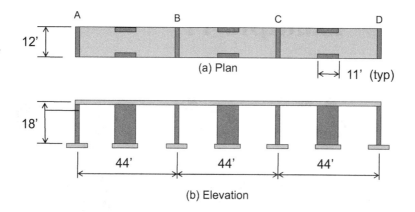

performed to determine if the diaphragm is flexible or semirigid. According to Section 12.3.1.3 and the related Fig. 12.3-1, the diaphragm is flexible if the maximum diaphragm deflection (MDD) is greater than 2.0 times the average drift of vertical element (ADVE).

An analysis to determine the quantities MDD and ADVE was performed using the SAP 2000 (CSI 2009) finite element analysis program. Thin-shell elements were used to model the walls and diaphragms. These elements automatically include in-plane shear deformations, which are essential for diaphragm analysis. **Fig. G11-2** shows the finite element model. Loading consisted of a 10-kip force applied in the Y direction at each node along the edge of the diaphragm. A uniform load applied to the edge of the diaphragm may also be used and would be more appropriate where the shell elements are not of a consistent width. A distributed load has been used in lieu of a concentrated load (as implied by Fig. 12.3-1) because the diaphragm's inertial forces would not be developed within a single point in an actual diaphragm. The use of a uniform load in lieu of a concentrated load also lowers the likelihood that the diaphragm would be classified as flexible.

Fig. G11-3 shows the deflections computed along the edge of the diaphragm. In the end span, the average drift of the vertical element (Fig. 12.3-1) is

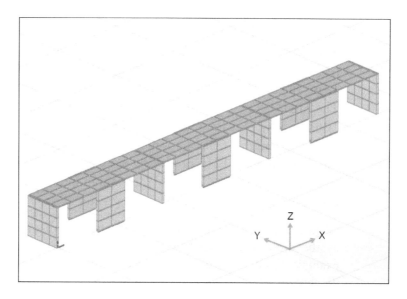

Fig. G11-3
Diaphragm displacements for a 4.0-in.-thick slab

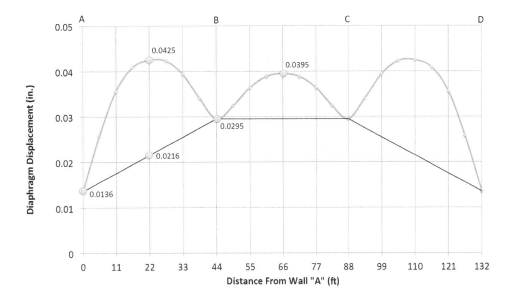

$$ADVE = (0.0136 + 0.0295)/2 = 0.0216 \text{ in.}$$

The maximum diaphragm deflection, MDD, is

$$MDD = 0.0425 - 0.0216 = 0.0209 \text{ in.}$$

The ratio of the maximum to the average displacement is

$$MDD/ADVE = 0.0209/0.0216 = 0.967$$

This ratio, 0.967, is less than 2.0, so according to Section 12.3.1.3, the diaphragm may not be considered flexible.

With a 10-kip force applied at each edge node, the total applied load on the structure is 250 kip. **Fig. G11-4** shows the distribution of these forces to the interior and exterior walls for various assumptions. For the computed assumption, based on the finite element analysis, each exterior wall resists 38.4 kip, and the interior walls carry 83.0 kip each, for a total (all four walls) of 242.8 kip. This total is less than 250 kip because the transverse walls carry

Fig. G11-4
Distribution of forces in walls for diaphragm thickness of 4.0 in.

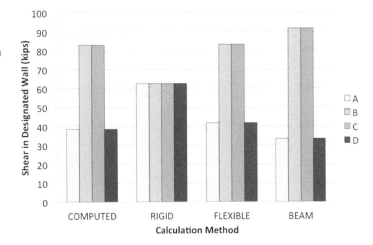

some shear, which is delivered to the foundation through weak-axis bending in these walls. For a fully rigid diaphragm, the 250-kip force would be expected to be equally distributed to the walls because the walls have the same lateral stiffness. This situation would result in a force of 62.5 kip in each wall. For a fully flexible diaphragm, the distribution of forces would be distributed on a tributary area basis with 1/6 of the total force, or 41.7 kip, going to the exterior walls, and 1/3 of the force, or 83.3 kip, going to each of the interior walls.

The finite element results appear to be more consistent with the flexible diaphragm assumption than with the rigid diaphragm assumption. This has nothing to do with diaphragm flexibility. Because the walls have virtually no torsional stiffness about the vertical axis, these supports emulate the condition of a pinned support more than a fixed support in a three-span continuous beam. For example, a three-span continuous beam subjected to the same loading as that used in the finite element analysis would have exterior support reactions of 33.3 kip and interior reactions of 91.7 kip. This result, shown as the beam assumption in **Fig. G11-4**, is valid regardless of the stiffness of the beam, as long as each span has the same flexural stiffness. Therefore, a prudent designer should consider how sensitive the results are to the assumptions made. Other parameters, such as cracking in walls and diaphragms, can influence the force distribution in the lateral system. If the results are sensitive to various parameters, the design should be based on a bounded solution.

If the diaphragm is assumed to be only 2.0 in. thick (e.g., concrete over a metal deck), the results of the finite analysis still indicate that the diaphragm is semirigid. The deflected shape for this condition is provided in **Fig. G11-5**. The computations are as follows:

$$\text{ADVE} = (0.0134 + 0.0297)/2 = 0.0215 \text{ in.}$$

Fig. G11-5
Diaphragm displacements for a 2.0-in.-thick slab

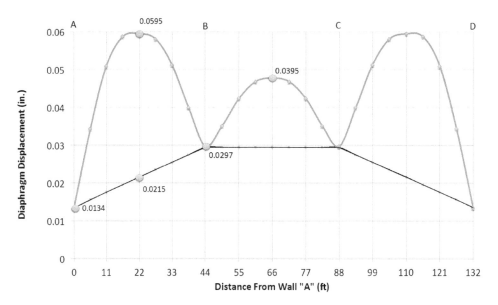

The maximum diaphragm deflection, MDD, is

$$\text{MDD} = 0.0595 - 0.0215 = 0.0380 \text{ in.}$$

The ratio of the maximum to the average displacement is

$$\text{MDD/ADVE} = 0.0380/0.0215 = 1.767$$

This value is less than 2.0, so even the system with the 2.0-in. slab is classified as semirigid.

The computed reactions at the base of the walls are close to those determined for the system with the 4.0-in.-thick diaphragm. Again, this example shows that diaphragm flexibility has little influence on how the shears are distributed. However, this behavior is not necessarily applicable to all buildings.

Accidental Torsion in Systems with Semirigid Diaphragms

Section 12.7.3 of ASCE 7 states that structures with semirigid diaphragms must be modeled to include the representation of the diaphragm stiffness (flexibility) and that additional degrees of freedom, aside from the two lateral displacements and one rotation at each level, must be included in the model. Clearly, any model that includes these effects is three-dimensional, so the de facto requirement of ASCE 7 is that systems with semirigid diaphragms must be modeled in three dimensions and presumably using a finite element approach wherein the diaphragm is discretized into a number of shell elements, as shown in **Fig. G11-2**. The question now arises as to whether the equivalent lateral force (ELF) method of analysis may be used to analyze the structure. Table 12.6-1 is silent on this issue because the only factor that excludes the possibility of using ELF, in Seismic Design Categories D through F, is a fundamental period greater than 3.5 T_S, or a system with a structural height greater than 160 ft and a horizontal irregularity of Type 1a or 1b or a vertical irregularity of Types 1a, 1b, 2, or 3.

In the author's opinion, the use of ELF for structures with semirigid diaphragms is not always appropriate, particularly if the diaphragm is somewhat flexible or highly irregular in shape. In ELF analysis, the lateral loads are applied at the center of mass, and if the diaphragms are modeled using shell elements, considerable local deformations and stress concentrations occur at the location of the applied load. The deformation and stress patterns in the actual diaphragm are quite different because the inertial forces in the diaphragm are distributed throughout the diaphragm, not at a single point. Some improvement would be obtained if the ELF story forces were applied to the diaphragms in a distributed manner. However, what this pattern should be is not clear because the pattern depends on the total accelerations at each point in the diaphragm and this pattern is unknown at the beginning of the analysis. A modal response spectrum analysis or a modal response history analysis produces realistic distributions of inertial forces (if the diaphragm masses are distributed throughout the diaphragm and if a sufficient number of modes are used in the analysis) and is therefore more appropriate than ELF.

Example 12

Structural Analysis Requirements

This example covers the selection of the structural analysis procedure (Section 12.6) and requirements for modeling the structure (Section 12.7).

Selection of Structural Analysis Procedure

Section 12.6 and Table 12.6-1 provide the requirements for selection of the structural analysis procedure to be used to determine displacements, drifts, and member forces caused by seismic load effects. Three basic procedures are provided in Table 12.6-1:

1. Equivalent lateral force (ELF) analysis,
2. Modal response spectrum (MRS) analysis, and
3. Linear or nonlinear response history (LRH or NRH) analysis.

Under certain circumstances, the simplified procedure provided in Section 12.14 may be used.

Table 12.6-1 shows that MRS and LRH (or NRH) analysis may be used for any system, and the ELF method may be used for almost any system. In fact, the essence of Table 12.6-1 can be restated as follows: The equivalent lateral force method of analysis may be used for any system, with the following exceptions:

1. Structures in Seismic Design Categories D, E, or F with structural height $h \leq 160$ ft and $T < 3.5T_S$ with any of the following irregularities:
    a. Type 1 horizontal irregularity (torsional or extreme torsional),
    b. Type 1 vertical irregularity (soft story and extreme soft story),
    c. Type 2 vertical irregularity (weight or mass), and
    d. Type 3 vertical irregularity (vertical geometric);

2. Structures in SDC D, E, or F with $h > 160$ ft and $T < 3.5T_S$ having any structural irregularities in Tables 12.3-1 or 12.3-2; and

3. Structures in Seismic Design Categories D, E, or F with $h > 160$ ft and $T \geq 3.5T_S$.

In terms of pure practicality, the ELF method of analysis should be used whenever permitted. Although the MRS method generally produces more accurate results than ELF, these results come at the expense of losing the signs (positive or negative) of seismic displacements and member forces, thereby complicating the combination of lateral seismic and gravity effects. The use of LRH analysis eliminates the problem with signs but requires much more effort on the part of the analyst, particularly as related to selection and scaling of ground motions. (See Example 6 in this guide for a description and discussion of the scaling process.) The complexity of NRH analysis is such that it should not be used except for special or important structures.

A potential advantage of MRS over ELF occurs when the computed system period of vibration is significantly greater than the upper limit period C_uT_a. This upper limit, described in Section 12.8.2, is used to provide a lower bound on the design base shear that is used in ELF, even when the computed period is greater than C_uT_a. When MRS analysis is used, Section 12.9.4 requires that the results be scaled such that the design base shear is not less than 85% of the base shear determined using ELF and the upper limit period. Therefore, the system analyzed with MRS could be designed for 15% less base shear than the same system designed using ELF.

It is also important to note that some form of ELF analysis is required in the analysis and design of all building structures. For example, ELF is almost certainly used in preliminary design, where checks for torsional irregularities and soft story irregularities require the development of static forces along the height of the building. Additionally, diaphragm forces (Section 12.10 and Equation 12.10-1) are based on ELF story forces.

Examples for Computing T_S and $3.5T_S$

The purpose of this example is to determine those circumstances under which it is likely that $T > 3.5T_S$, thereby disallowing the use of ELF for structures that have been assigned to Seismic Design Categories D, E, and F. The example is based on a site with mapped MCE spectral accelerations $S_s = 0.8\,g$ and $S_1 = 0.25\,g$. Soil Site Classes B, C, D, and E are also considered. The quantities T_S and $3.5T_S$ and the height and approximate number of stories of steel moment frame and braced frame buildings that would have a period of $3.5\,T_s$ are determined. Heights are based on the following:

$$T = 3.5T_S = C_uT_a \qquad \text{(Section 12.8.2)}$$

where $C_u = 1.4$ from Table 12.8-1

$$T_a = C_t h_n^x \qquad \text{(Eq. 12.8-7)}$$

Table G12-1 T_s Values for SDC D Structures on Various Sites

Site	F_a	F_v	S_{DS} (g)	S_{D1} (g)	T_s (s)	$3.5T_s$(s)	h_n M.F. (ft) (No. of Stories)	h_n B.F. (ft) (No. of Stories)
B	1.00	1.00	0.533	0.167	0.313	1.094	97 (8)	207 (16)
C	1.08	1.55	0.576	0.258	0.448	1.570	153 (12)	336 (26)
D	1.18	1.90	0.629	0.317	0.503	1.76	177 (14)	359 (28)
E	1.14	3.00	0.608	0.500	0.822	2.878	327 (26)	754 (60)

Note: Table is based on $S_s = 0.8\,g$ and $S_1 = 0.25\,g$. M.F. = moment frame; B.F. = braced frame.

According to Table 12.8-2, C_t and x are, respectively, 0.028 and 0.8 for steel moment-resisting frames, and 0.02 and 0.75, respectively, for concentrically braced frames.

T_S is defined in Section 11.4.5 and is computed as follows:

$$T_S = \frac{S_{D1}}{S_{DS}}$$

Physically, T_S is the period at which the constant acceleration [Eq. (12.8-3)] and the constant velocity [Eq. (12.8-4)] branches of the response spectrum meet (Fig. 11.4-1).

The results of the calculations are presented in **Table G12-1**. The T_S values are given in column 6, and $3.5T_S$ values in column 7. Clearly, the $3.5T_S$ values increase with increasing softness of the soil. This phenomenon occurs because the site amplification factor F_v is always greater than or equal to the factor F_a.

Columns 8 and 9 of **Table G12-1** show the heights of the structures in feet that give periods equal to $3.5T_S$. Also shown in parentheses is the number of stories, assuming that all stories have a height of 12.5 ft. The values in column 9, for the braced frame, are particularly interesting because the heights for all site classes are greater than the height limit of 160 ft given in Table 12.2-1 for a special concentrically braced frame. The conclusion that may be drawn from this is that the ELF method of analysis may be used for all regular concentrically braced frames (except for some structures in Site Class B, taller than 207 ft, which have height limits extended to 240 ft as allowed by Section 12.2.5.4). Regarding regular special moment frames, the values in column 8 of **Table G12-1** indicate that the ELF method may be used for any regular structure that is less than approximately eight stories high.

Structural Analysis Considerations

Aside from the method of analysis, the other principal modeling decisions to make are the modeling of roof and floor diaphragms and whether three-dimensional analysis is required.

Section 12.7.3 requires that 3D modeling be used when horizontal structural irregularities of Types 1, 4, or 5 of Table 12.3-1 exist. When a 3D model is required, the diaphragms may be modeled as rigid only if the diaphragms are rigid in accordance with Section 12.3.1.2. Otherwise, the diaphragms must be modeled as semirigid.

Section 12.7.3 provides several other modeling requirements, such as inclusion of P-delta effects, representation of strength and stiffness of nonstructural components, inclusion of cracking in concrete and masonry structures, and modeling of deformations in the panel zone region of steel moment frames. A brief discussion of these issues is provided in Example 19. However, considerable thought and experience are required to develop analytical models for structural analysis. The more advanced the analysis, the more detail required in the model and the more time and thought required. Additionally, validation of complex models is much more difficult than it is for simpler models.

Example 13

Use of the Redundancy Factor

Section 12.3.4 describes the methodology for determination of the redundancy factor, ρ, which is used in many of the seismic load combinations that are specified in Section 12.4. This example demonstrates how the redundancy factor is determined for several structural systems. The use of the redundancy factor in the context of the load combinations is demonstrated in Example 18 of this guide.

The redundancy factor, ρ, is used in association with the seismic load combinations that are specified in Section 12.4. In particular, in accordance with Eq. (12.4-3):

$$E_h = \rho Q_E \qquad \text{(Eq. 12.4-3)}$$

where E_h is the effect of horizontal seismic forces, and Q_E is a component or connection force that results from application of horizontal loads. When used in conjunction with Eq. (12.4-3), the value of the redundancy factor is either 1.0 or 1.3, depending on the Seismic Design Category (SDC) and the structural configuration. The factor can be different in the two horizontal directions. Where the redundancy factor is determined to be 1.3 in one direction, it applies to all connections and components in the structure that are designed to resist seismic loads in that direction. Section 12.3.4.1 lists existing exceptions. For example, ρ can be taken as 1.0 in the design of nonstructural components. The factor ρ can also be taken as 1.0 in drift and P-delta calculations. Note, however, that the allowable story drifts must be divided by ρ for moment frames in SDC D, E, and F (see Section 12.12.1.1).

For structures in SDCs B and C, ρ is taken as 1.0 in each direction. For structures in SDCs D, E, and F, the redundancy factor is 1.3, but may be reduced to 1.0 if it "passes" *either* a configuration test or a calculation test. These tests are stipulated in Section 12.3.4.2.

The configuration test is described in subparagraph (b) of Section 12.3.4.2, which states that ρ may be taken as 1.0 where the structure has no horizontal structural irregularities and where at least two bays of perimeter seismic force–resisting elements exist on each side of the building for each story of the building resisting more than 35% of the seismic base shear. The number of bays for a shear wall shall be considered as the total length of the wall (in one plane) divided by the story height, or two times the total length of the wall divided by the story height for light-frame construction.

Subparagraph (a) of Section 12.3.4.2, in association with Table 12.3-3, describes the calculation test. In this test, a lateral load–resisting element (or connection) is removed from the structure to determine if removal of the element or connection causes an extreme torsional irregularity (where one was not present before), or if the lateral strength of the structure is reduced by more than 33%. If the extreme torsional irregularity or excessive strength loss does not occur, the redundancy factor may be taken as 1.0. Note that the torsional irregularity test fails (ρ must be taken as 1.3) if the structure has an extreme torsional irregularity before the component or connection is removed.

As stated earlier, ρ = 1.0 may be used in SDC D, E, and F structures if either condition (a) or (b) applies. The condition (a) test is much more difficult to apply, so test (b) should be applied first. It is also noted that the loss-of-strength test under condition (a) rarely occurs and that providing a logical argument (performing the "calculation" by inspection) that a 33% strength loss is impossible for the given configuration is generally acceptable. The torsion test is more problematic, particularly when the structure has a torsional irregularity before the component or connection is removed.

Fig. G13-1 illustrates several cases for which condition (b) may be evaluated. The floor plans in the figure are applicable at levels for which the

Fig. G13-1
Evaluation of the redundancy factor for various buildings

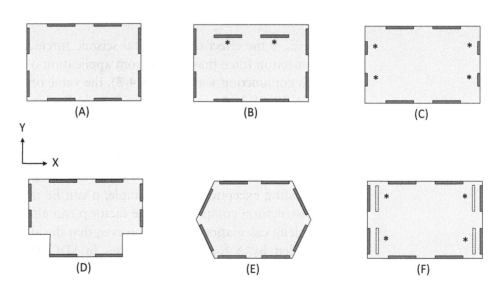

seismic base shear is greater than 35% of the base shear. Only building A satisfies the condition (b) test. The building has no horizontal structural irregularities, the walls on each side of building are long enough to provide two equivalent bays on each side, and the walls are located on the perimeter. Building B violates the criteria because the two walls marked with asterisks are not on the perimeter. Building C, which is assumed to have no irregularities, does not satisfy the criteria in the Y direction because the plan length of the walls marked with asterisks is insufficient to provide two equivalent bays on each side of the building. Buildings D, E, and F cannot automatically be classified with ρ = 1.0 because each has a horizontal structural irregularity. In building F, the irregularity occurs because of an out-of-plane offset of the shear walls marked with asterisks. The walls at the upper level are on the interior of the building and transfer to the exterior at the lower levels.

The fact that condition (b) has not been satisfied for a given building does not mean that the redundancy factor is 1.3. This situation would be the case only if condition (a) in Section 12.3.4.2 is also not met. Consider again building B of **Fig. G13-1**. In this shear wall system, each wall has a plan length greater than the height of the wall. Thus, the height-to-width ratio of the walls is less than 1.0, and the system defaults to "other" lateral force–resisting elements in Table 12.3-3. Presumably, therefore, this system can be assigned a redundancy factor of 1.0 in each direction because no requirements dictate otherwise. The same situation appears to occur even if the walls marked by asterisks in building B of **Fig. G13-1** were removed entirely. In the opinion of the author, this situation violates the spirit of the redundancy factor concept, and a factor of 1.3 should be assigned in this case.

In building C of **Fig. G13-1**, each wall marked with an asterisk has a length less than the story height. Removal of one of these walls does not cause an extreme torsional irregularity. At first glance the removal of one wall appears to reduce the strength of the system by only 25% in the Y direction. However, this situation does not consider the effect of torsion. The reduction in strength must be based on the questions, "How much lateral load can be applied in the Y direction for the system with one wall missing, and how does that compare with the strength of the system with the wall in place?"

Two interpretations exist for evaluation of the strength of the system with elements removed. The first is based on elastic analysis, and the second is based on inelastic analysis. An important consideration of the use of an inelastic analysis is that the system must be sufficiently ductile to handle the continued application of loads after the lateral load–resisting elements begin to yield.

Consider, for example, the system shown in **Fig. G13-2**. This system has eight identical walls, marked A through H, each with a force–deformation relationship as shown in **Fig. G13-3**. The lateral load–carrying capacity of each wall is 100 kip.

The system is evaluated on the basis of the following situations:
1. Elastic behavior with all walls in place,
2. Elastic behavior with one wall removed,

Fig. G13-2

System with one wall (C) removed

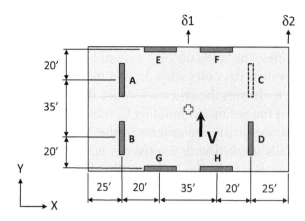

Fig. G13-3

Force-deformation relationship for shear wall

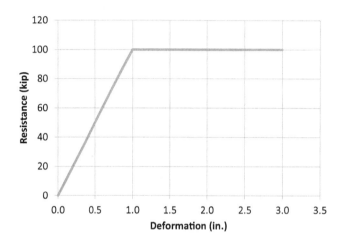

3. Inelastic behavior analysis with all walls in place, and
4. Inelastic behavior with one wall removed.

For each situation, the lateral force V is applied in the y direction at an eccentricity of 5% of the width (eccentricity = 0.05×125 ft = 6.25 ft) of the building in the x direction. The eccentricity of 5% of the plan width is consistent with the accidental torsion requirements of Section 12.8.4.2.

The analysis was performed for this example using a computer program that can model inelastic structures. This analysis provided the curves shown in **Fig. G13-4**. The upper curve represents the behavior of the system with all four walls in place, and the lower curve is for the system with wall C removed. The elastic analysis for each system is represented by the response up to first yield (the first change in slope of the curves), and the inelastic response is represented by the full curve.

From the perspective of the elastic analysis, the structure with all walls in place can resist a lateral load V of 370 kip. At this load, walls C and D on the right side of the building carry 100 kip, and walls A and B carry 85 kip. The ratio of the displacement $\delta 2$ relative to the displacement at point $\delta 1$ is 1.14, so according to Table 12.3-1, the structure does not have a torsional irregularity.

When the inelastic response is considered, the structure can carry additional lateral load because walls A and B can each resist an additional 15 kip before they reach their 100-kip capacity. With four walls resisting 100 kip

Fig. G13-4

Force-deformation plot for structure with three or four walls using inelastic analysis

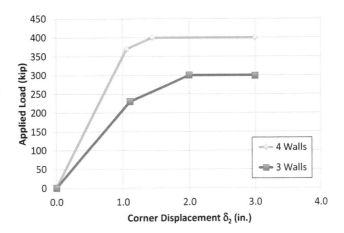

each, the total lateral capacity of the system is 400 kip. Recall that the force-displacement plot for this four-wall system is shown by the upper curve in **Fig. G13-4**. The displacement shown in the figure is the y direction displacement $\delta 2$. The first change of slope in the curve occurs when walls C and D yield, and the second change occurs when walls A and B yield.

When wall C is removed, the center of rigidity moves 9.375 ft to the left. When an elastic analysis is performed, the system can resist a lateral load of only 230 kip because wall D reaches its 100-kip capacity. At this point, walls A and B resist 65 kip each. Additionally, the ratio of the displacement at point $\delta 2$ with respect to the displacement at point $\delta 1$ is 1.35. Hence, a torsional irregularity, but not an extreme irregularity, exists. The ratio of the resisting force of the three-wall system to that of the four-wall system is 230/370 or 0.621. On this basis, the system must be designed with $\rho = 1.3$ because it loses more than 33% of its strength.

From the perspective of an inelastic analysis, the three-wall system can resist a total of 300 kip. This is shown by the force-displacement diagram (the lower curve) in **Fig. G13-4**. The ratio of the inelastic resisting force of the three-wall system to that of the four-wall system is 300/400 = 0.75. In this case, the system could theoretically be designed with $\rho = 1.0$ because it passes both the strength and the nonextreme torsion irregularity tests.

Two final points are made regarding the use of the redundancy factor:

1. When it is determined that $\rho = 1.3$ in a given direction, the factor of 1.3 applies only to load combinations where seismic forces are applied in that direction. Also, the load combination with $\rho = 1.3$ is used for all components and connections developing seismic forces when the load is applied in that direction. This includes elements and component from the bottom to the top of the structure and is *not* limited to those elements and components in levels resisting more than 35% of the seismic base shear.

2. The condition (a) test of Section 12.3.4.2 applies for buildings with different types of lateral load–resisting elements in a given direction. For example, for a dual moment-frame shear wall system, the test would be performed with a single wall removed and then with a single beam removed from the moment frame.

Example 14

Accidental Torsion and Amplification of Accidental Torsion

This example considers several issues related to torsional loading. Included are torsional irregularities, accidental torsion, torsional amplification, and application of accidental torsion to structures analyzed using the equivalent lateral force or modal response spectrum approaches. Systems with rigid and semirigid diaphragms are considered. Accidental torsion need not be considered for systems with flexible diaphragms.

Fig. G14-1 is a plan view of a five-story reinforced concrete shear wall building. The first story is 12 ft, 4 in. tall, and the upper stories are each 11 ft, 4 in. tall. The building, located in central Missouri, houses various business offices and is classified as Risk Category II. The following design spectral accelerations have been determined for the site:

$S_{DS} = 0.45 \, g$, and
$S_{D1} = 0.19 \, g$.

Tables 11.6-1 and 11.6-2 indicate that the Seismic Design Category is C. Table 12.2-1 allows the use of an ordinary reinforced concrete shear wall, with the following design parameters:

$R = 5$, and
$C_d = 4.5$.

Fig. G14-1

Plan of reinforced
concrete shear wall
building

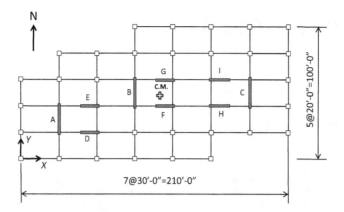

For this analysis, all walls are assumed to be 10 in. thick and constructed with 4,000 lb/in.^{2} normal weight concrete. Walls A, B, and C have a length of 22 ft, and walls D, E, F, G, H, and I have a length of 16 ft. Floor and roof diaphragms are assumed to be 6.0-in. thick solid slabs, constructed with 4,000 lb/in.^{2} normal weight reinforced concrete.

The period of vibration is estimated from Eq. (12.8-7). Using $h_n = 12.33 + 4(11.33) = 57.7$ ft and coefficients $C_t = 0.02$ and $x = 0.75$ from Table 12.8-2,

$$T = T_a = C_t h_n^x = 0.02 \times 57.7^{0.75} = 0.42 \text{ s}$$

This period is used for analysis (in lieu of $C_u T_a$) because an analytical period (from a computer program) is not available. It is noted, however, that assessment of torsional irregularity and torsional amplification is not strongly period dependent.

Lateral forces are computed using the equivalent force method. Using the ground motion parameters given in the aforementioned, $T_S = S_{D1}/S_{DS} = 0.19/0.45 = 0.42$, which by coincidence is equal to $T = 0.42$ s, so Eqs. (12.8-2) and (12.8-3) produce the same base shear. Using a total seismic weight of the building of 9,225 kip, the base shear is determined from Eq. (12.8-2):

$$C_s = \frac{S_{DS}}{\left(\dfrac{R}{I}\right)} = \frac{0.45}{\left(\dfrac{5.0}{1.0}\right)} = 0.090$$

$$V = C_s W = 0.090(9,225) = 830 \text{ kip}$$

Equivalent lateral forces are computed in accordance with Section 12.8.3, with $k = 1.0$. The results of the calculation are provided in **Table G14-1**.

Due to the plan shape (see **Fig. G14-1**) the structure has a reentrant corner irregularity. Hence, according to Section 12.3.1.2, the diaphragm may not be classified as rigid and must be analyzed as semirigid (as it is clearly not a flexible diaphragm). By the requirement of Sections 12.8.4.1 and 12.8.4.2, inherent torsion and accidental torsion must be included in the analysis. A three-dimensional analysis is required for the structure because the in-plane deformations of the semirigid diaphragm must be

Table G14-1   Equivalent Lateral Forces

Level	H *(ft)*	h *(ft)*	W *(kip)*	Wh^k	Wh^k/*Total*	F *(kip)*
5	11.33	57.66	1,820	104,941	0.326	271
4	11.33	46.33	1,845	85,479	0.266	221
3	11.33	35.00	1,845	64,575	0.201	167
2	11.33	23.66	1,845	43,653	0.136	113
1	12.33	12.33	1,870	23,057	0.070	58
Total	57.65	—	9,225	321,705	1.00	830

included (see Section 12.7.3), and this cannot be reasonably done without a three-dimensional analytical model. There is no need to compute separately the effects of inherent torsion because such effects are automatically included in a three-dimensional analysis.

A 3D analysis is also needed to determine if a torsional irregularity exists and whether the accidental torsion must be magnified. For this example such an analysis was run using a 3D finite element analysis program, wherein the walls were modeled as membrane elements. Membrane elements were used in lieu of shell elements because excluding the out-of-plane stiffness of the walls was desirable for simplicity. The use of membrane elements automatically includes in-plane axial, flexural, and shear deformation in the walls. Uncracked properties were used because the main purpose of the analysis is to determine the elastic deformations in the system and to determine the distribution of the forces in the walls. Although cracking affects the absolute magnitude of displacements, it does not affect the ratio of the displacements at the edge of the building to the displacement at the center of the building, as long as the same stiffness reduction factors are used to represent cracking in each wall. Similarly, cracking does not affect the distribution of forces if all walls are cracked to the same degree. (This discussion is based on flexural properties and flexural cracking. For shear wall systems, the effect of shear deformations and shear cracking should also be considered because these effects can have a significant influence on the distribution of forces in the system.)

To determine the possible presence of a torsional irregularity, the concrete diaphragms were modeled with 6.0-in. thick membrane elements. The concrete was assumed to be uncracked.

The analysis is carried out only for forces acting in the north–south direction. Three load conditions are applied, one without accidental eccentricity, one with the lateral force applied east of the center of mass, and the other with the forces applied west of the center of mass. The location of the center of mass is shown in **Fig. G14-2**.

As required by Section 12.8.4.2, the lateral forces are applied at an eccentricity of 0.05 times the length of the building perpendicular to the direction of loads. Thus the eccentricity is 0.05(210) = 10.5 ft when the lateral loads are applied in the north–south direction. Because of the rigid

Fig. G14-2

Plan shown with loading and monitoring locations. C.M. is center of mass

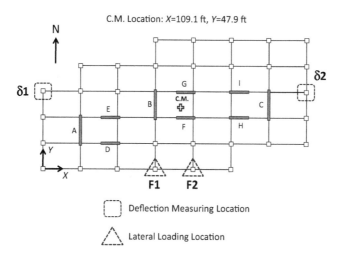

C.M. Location: X=109.1 ft, Y=47.9 ft

Deflection Measuring Location

Lateral Loading Location

diaphragm assumption, the lateral load, including torsion, may be applied through the use of two loading points, shown as F1 and F2 in **Fig. G14-2**. (The loads could also be applied as a single concentrated force [lateral load] plus moment applied about the vertical axis [accidental torsion]. These loads would be applied at a single node at the center of mass.) The forces caused by the lateral load without torsion are based on simple beam reactions for a beam with a span of 30 ft, as shown in **Fig. G14-3(a)**. The accidental torsional component of load is also applied as two concentrated forces, shown in **Fig. G14-3(b)**. The total load is simply the lateral load, plus or minus the torsional load, as illustrated in **Figs. G14-3(c)** and **G14-3(d)**.

To determine if a torsional irregularity exists (Table 12.3-1), interstory drifts (not displacements) are monitored at the extreme edges of the building under a loading that consists of the design lateral forces and (plus or minus) accidental torsion. If the maximum drift at the edge of any story exceeds 1.2 times the average of the drifts at the two edges of the story, a torsional

Fig. G14-3

Loading values applied to building

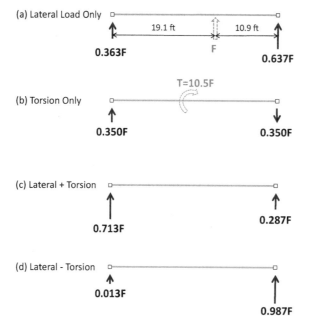

irregularity exists. If the maximum drift exceeds 1.4 times the average, an extreme irregularity exists. In this example, the computed displacements were used without the deflection amplifier C_d because the C_d term cancels out when computing the ratio of drifts.

For determining the torsional amplification factor, the story displacements (not drifts) are used. An amplification factor, A_x, is then computed for each story. This factor is determined using Eq. (12.8-14):

$$A_x = \left(\frac{\delta_{max}}{1.2\delta_{avg}} \right)^2$$

where δ_{max} is the maximum deflection at the edge of the story, and δ_{avg} is the average deflection at the two edges. The two points at which the deflections were monitored are designated as $\delta 1$ and $\delta 2$ in **Fig. G14-2**.

Inherent torsion must be included in the analysis, which is automatically accomplished when a 3D analysis is performed. Separating out the inherent torsion is not needed, however, because it is never used independently. Nevertheless, applying the lateral loads without accidental torsion and observing the resulting deflection patterns is worthwhile. These deflections indicate the location of the center of rigidity relative to the center of mass. Accidental torsion loadings that rotate the floor plates in the same direction as the rotation resulting from inherent torsion clearly control when determining if torsional irregularities occur and when computing amplification factors.

To determine the displacements in the system under lateral load only, the forces listed in column 7 of **Table G14-1** were applied as shown in **Fig. G14-3(a)**. The results of the analysis are presented in **Table G14-2**. The rotational measurement, θ, is the rotation about the vertical axis, which is counterclockwise positive. The deflections at point $\delta 2$ at any level are always greater than those at $\delta 1$, causing the building to twist counterclockwise, which is a positive rotation. This situation indicates that the center of mass lies east of the center of rigidity, which is to the right in **Fig. G14-2**. Clearly, lateral loads plus torsion results in greater twisting in the positive direction.

The results in **Table G14-3** indicate an increasing rotation when the torsion is applied in the positive direction. The twisting causes an extreme torsional irregularity. A given irregularity type needs to occur only at one story for the whole building to be classified as having that irregularity.

Table G14-2 Displacements for Building under Lateral Force without Torsion

Level	δ1 (in.)	δ2 (in.)	θ (radians)
5	0.463	0.754	1.15E–04
4	0.341	0.555	8.50E–05
3	0.224	0.364	5.56E–05
2	0.120	0.196	2.98E–05
1	0.042	0.068	1.04E–05

Table G14-3   Results for Lateral plus Accidental Torsion: Irregularity Check

Level (δ) Story (Δ)	δ1 (in.)	δ2 (in.)	Δ1 (in.)	Δ2 (in.)	Δ_{avg} (in.)	$\Delta_{max}/\Delta_{avg}$	Torsional Irregularity
5	0.304	0.929	0.080	0.245	0.163	1.508	Extreme
4	0.224	0.684	0.077	0.235	0.156	1.510	Extreme
3	0.147	0.449	0.068	0.208	0.138	1.505	Extreme
2	0.079	0.241	0.052	0.157	0.104	1.507	Extreme
1	0.027	0.084	0.027	0.084	0.056	1.506	Extreme

Table G14-4   Results for Lateral plus Accidental Torsion: Amplification Factors

Level (δ)	δ1 (in.)	δ2 (in.)	δ_{avg} (in.)	$\delta_{max}/\delta_{avg}$	A_x
5	0.304	0.929	0.617	1.508	1.578
4	0.224	0.684	0.454	1.507	1.578
3	0.147	0.499	0.298	1.507	1.577
2	0.079	0.241	0.160	1.507	1.577
1	0.027	0.084	0.056	1.506	1.576

Table G14-5   Results for Lateral minus Accidental Torsion: Irregularity Check

Level (δ) Story (Δ)	δ1 (in.)	δ2 (in.)	Δ1 (in.)	Δ2 (in.)	Δ_{avg} (in.)	$\Delta_{max}/\Delta_{avg}$	Torsional Irregularity
5	0.623	0.579	0.165	0.153	0.159	1.038	None
4	0.458	0.426	0.157	0.146	0.152	1.036	None
3	0.301	0.280	0.139	0.130	0.135	1.033	None
2	0.162	0.150	0.106	0.098	0.102	1.039	None
1	0.056	0.052	0.056	0.052	0.054	1.037	None

Table G14-4 shows the computation of the torsional amplification factor at each level. In this case, the values are virtually the same all the way up the building because the walls continue the full height of the building.

When the torsion is applied in the opposite direction, it offsets the inherent torsion and the irregularity disappears. Additionally, the computed amplification factors are less than 1.0, so the minimum factor of 1.0 is applied. If the torsional irregularity occurs for any direction of the applied eccentricity, the system is classified as having a torsional irregularity. These results are provided in **Tables G14-5** and **G14-6**.

A few points are worthy of discussion at this point:
1. When determining torsional amplification, it is not necessary to iterate by analyzing the system with the amplified accidental torsion, determining a new amplification factor, analyzing again, and so on. A single analysis using the 5% accidental eccentricity determines amplification factors.

Table G14-6   Results for Lateral minus Accidental Torsion: Amplification Factors

Level (δ)	δ1 (in.)	δ2 (in.)	δ_{avg} (in.)	$\delta_{max}/\delta_{avg}$	A_x
5	0.623	0.579	0.601	1.037	1.0
4	0.458	0.426	0.442	1.036	1.0
3	0.301	0.280	0.291	1.036	1.0
2	0.162	0.150	0.156	1.038	1.0
1	0.056	0.052	0.054	1.037	1.0

Note: Minimum $A_x = 1.0$.

2. When assessing torsional regularity or torsional amplification, applying the accidental torsion simultaneously in the two orthogonal directions is not necessary.

Application of Accidental Torsion in Systems with Relatively Flexible Semirigid Diaphragms

Earlier this example noted that the floor and roof diaphragms were modeled with 6.0-in. thick membrane elements. Although the diaphragm is classified as semirigid, it will behave like a rigid diaphragm. In some structures, a diaphragm that is classified as semirigid may behave more like a flexible diaphragm. In such cases the lateral forces should not be applied as shown in **Fig. G14-2**. Instead, these forces should be distributed throughout the diaphragm in some reasonable manner. A possible distributed loading for the structure analyzed in this example is shown in **Fig. G14-4**. In this figure, the lateral force at a level is applied on the basis of nodal forces, where the sum of the individual nodal forces is equal to the total load applied at the

Fig. G14-4
Direct and torsional loading for a system with a semirigid diaphragm

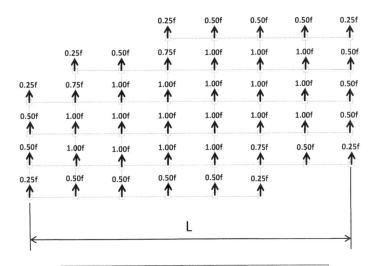

For accidental torsion apply a Z-direction moment of 0.05L times nodal force at each loading location.

level. Nodal forces are based on a tributary area (mass) basis. Accidental torsion is applied by a series of moments applied in the plane of the diaphragm. Shell, not membrane, elements would be required for such an analysis, and the element formulation used in the finite element analysis program must be able to accommodate drilling degrees of freedom (nodal moments applied about an axis normal to the plane of the diaphragm). If applying moments directly to the nodes is not possible, the torsion may be applied by modifying the lateral forces that are applied to each node.

It is important to note, however, that indiscriminate use of ELF analysis with semirigid diaphragms is not appropriate. This is particularly true when the diaphragm tends to be more flexible than rigid or when the diaphragm has a highly irregular shape (e.g., the diaphragms with large openings in Example 9, **Fig. G9-6**).

Application of Accidental Torsion and Torsional Amplification in Modal Response Spectrum Analysis

The loading shown in **Fig. G14-4** would be useful in determining if torsional irregularities exist, computing torsional amplification factors, and applying the lateral loading in an equivalent lateral force analysis. If a modal response spectrum (MRS) analysis is used, applying the accidental torsion in a truly dynamic manner (e.g., by physical adjustment of the center of mass) is difficult. Doing so results in a different set of modal properties (frequencies and mode shapes) for each adjustment of mass. Instead, it is recommended that the effects of accidental torsion be included in a separate static analysis and then combined with the results of the MRS analysis. A similar approach is recommended when a rigid diaphragm assumption is applicable.

When applying the accidental torsion statically, equivalent lateral forces, scaled to produce the same base shear as that produced by the MRS analysis, would be applied at a 5.0% eccentricity, exactly as is done in a full ELF analysis. Application of accidental torsion using an ELF force distribution is permitted even where an MRS analysis (or higher) is required by Table 12.6-1.

However, whenever static loading is used to apply the accidental torsion, it is necessary for torsionally irregular buildings in SDC C and above to amplify the accidental torsion according to the requirements of Section 12.8.4.3. This is required even when the main lateral load analysis is performed using MRS analysis because the accidental torsion effects are not included in the dynamic model per the requirements of Section 12.9.5. Accidental torsion would be included in the dynamic model only if the mass locations were physically adjusted.

Example 15

Load Combinations

This example explores the use of the strength design load combinations that include earthquake load effects. These load combinations, numbered 5 and 7 in Section 2.3.2 of ASCE 7, are then discussed in context with the requirements of Section 12.4. Also discussed in this example are requirements for including direction of loading (Section 12.5), accidental torsion (Section 12.8.4.2), and amplification of accidental torsion (Section 12.8.4.3).

Chapter 2 of ASCE 7 provides the required load combinations for both strength-based and allowable stress-based designs. This example covers only the use of the strength-based load combinations. Seven basic load combinations are provided in Section 2.3.2. Each member and connection of the structure must be designed for the maximum force or interaction of forces (e.g., axial force plus bending) produced by any one of these basic combinations. For any given member, such as a reinforced concrete girder, different combinations may be found to control different aspects of the design. For example, load combination 2 in Section 2.3.2 might control the requirements for bottom reinforcement at midspan, whereas combination 5 controls requirements for top reinforcement at the ends of the member. A specific example of this circumstance is provided later. The remainder of this example concentrates on combinations 5 and 7:

Combination 5:	$1.2D + 1.0E + 1.0L + 0.2S$, and
Combination 7:	$0.9D + 1.0E$.

In these combinations, the factor on earthquake load effects, E, is 1.0. This is due to the fact that the spectral design accelerations S_{DS} and S_{D1}, produced from the requirements of Chapter 11, are calibrated to be consistent with an ultimate load. The factor on live load in combination 5 may be reduced to 0.5 in most cases (see Exception 1 in Section 2.3.2). The adopted building code may provide load combinations that are somewhat different than those specified in ASCE 7. If so, these combinations must be used in lieu of ASCE 7 requirements. For example, the 2012 IBC (ICC 2011) specifies a factor of 0.7 on snow load in combination 5 for configurations that do not shed snow off the structure (such as saw tooth roofs).

The snow load in combination 5 is always included when $S > 0$. There might also be a snow load effect in E in both combinations because the effective seismic weight, W, is required to include 20% of the design snow load when the flat roof snow load exceeds $30\,\mathrm{lb/ft^2}$ (Section 12.7.2).

The combinations are used in two ways. The first, covered in Section 12.4.2, applies to all elements and connections in the structure and may be considered the standard load combinations. The second, covered in Section 12.4.3, is for those special elements or connections that must be designed with the overstrength factor, Ω_o. ASCE 7 provides several specific cases where the overstrength load combination must be used. For example:

- Section 12.2.5.2, which requires that the foundation and other elements contributing to the overturning resistance of cantilever column structures be designed with the overstrength factor;
- Section 12.3.3.3, which pertains to elements supporting discontinuous walls or frames;
- Section 12.10.2.1, which pertains to collector elements, their splices, and their connections to resisting elements; and
- Chapter 14, which triggers the use of materials standards such as Section 14.1.2.2 (referring to AISC 341) or Section 14.2 (referring to ACI 318 Chapter 21) and other materials standards that trigger the use of the overstrength factor in load combinations

Section 12.4.2 of ASCE 7 provides details on the standard seismic load effect. For use in load combination 5, the seismic load effect E is given as

$$E = E_h + E_v \qquad\qquad\text{(Eq. 12.4-1)}$$

and for use in load combination 7,

$$E = E_h - E_v \qquad\qquad\text{(Eq. 12.4-2)}$$

where

$$E_h = \rho Q_E \qquad\qquad\text{(Eq. 12.4-3)}$$

and

$$E_v = 0.2 S_{DS} D \qquad\qquad\text{(Eq. 12.4-4)}$$

The term E_h represents the horizontal seismic load effect. The term ρ in Eq. (12.4-3) is the redundancy factor, computed in accordance with Section 12.3.4. This value is 1.0 for all buildings assigned to Seismic Design Category (SDC) B or C and is either 1.0 or 1.3 in SDC D through F. This factor applies to the entire structure but may be different in the two orthogonal directions. See Example 13 in this guide for details on determination of the redundancy factor.

The term E_v represents the effect of vertical ground acceleration, which is not considered explicitly elsewhere, with the exception of Section 12.4.4, which provides requirements for minimum upward forces in horizontal cantilevers for buildings in SDC D through F.

Q_E in Eq. (12.4-3) is the seismic effect on an individual member or connection. Seismic analysis of the structure produces this value, which includes direct loading (e.g., application of equivalent lateral forces), accidental torsion and torsional amplification (if applicable), and orthogonal loading effects (if applicable). Q_E might represent, for example, a bending moment at a column support, an axial force in a bracing member, or a stress in a weld. In some cases, Q_E might represent an interaction effect, such as an axial-force bending moment combination in a beam column. In such cases, both the axial force and bending moment occur concurrently and should be taken from the same load combination.

When Eqs. (12.4-1) through (12.4-4) are substituted into the basic load combinations, the following detailed combinations for strength design are obtained:

Combination 5: $(1.2 + 0.2S_{DS})D + \rho Q_E + 1.0L + 0.2S$, and
Combination 7: $(0.9 - 0.2S_{DS})D + \rho Q_E$.

Figs. G15-1 and **G15-2** illustrate the use of these load combinations, showing a simple frame with gravity and seismic loading. Snow loading is not present.

The top of **Fig. G15-1** shows only the gravity portion of the load, with the heavy gravity case shown at the upper left and the light gravity case shown at the upper right. At the bottom of the figure, the loading is shown for seismic effect acting to the east or to the west. Moment diagrams are drawn for each loading, and these diagrams are presented on the tension side. The moment values (units not important) are shown for each loading.

Fig. G15-1
Basic load combinations for simple frame

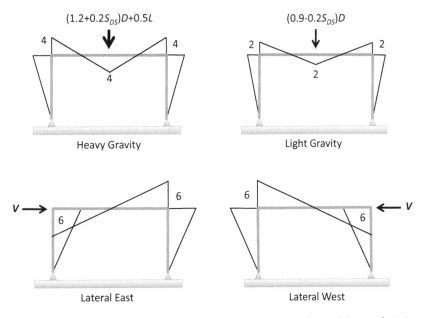

Fig. G15-2

Combinations of basic combinations for worst effect

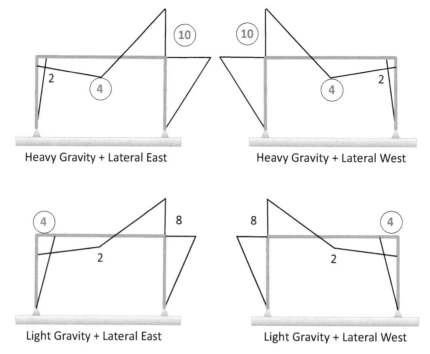

Heavy Gravity + Lateral East

Heavy Gravity + Lateral West

Light Gravity + Lateral East

Light Gravity + Lateral West

Fig. G15-2 shows the combination of gravity and earthquake load. The top of the figure gives the total moments for load combination 5, with heavy gravity plus seismic acting to the east on the left side of the figure and heavy gravity plus seismic acting to the west on the right side of the figure. The bottom of the figure is for combination 7 with light gravity and seismic acting to the east or west. Controlling moments are circled.

Fig. G15-2 shows that each load combination must be exercised twice, once for positive seismic and once for negative seismic to produce the controlling effect in each member or connection. For the given example, the controlling tension on the top moment is 10 at both ends of the beam, and the controlling tension on the bottom moment is 4 for the full beam span. Of course, other load combinations (without seismic) provided in Chapter 2 of ASCE 7 must also be exercised to determine if they control. It is important to recognize, however, that the seismic detailing requirements associated with any system must be provided, regardless of the loading combination that controls the strength of the member or connection. For example, a member in an intermediate moment frame that has a wind-based design force twice as high as the seismic design force would be sized on the basis of the wind forces but must be detailed according to the requirements for intermediate moment frames.

The lateral forces shown in **Figs. G15-1** and **G15-2** include the effects of accidental torsion and the effect of seismic loads acting simultaneously in orthogonal directions. Accidental torsion must be considered for any building with a nonflexible diaphragm, and torsionally irregular buildings in SDC C through F are subject to requirements for amplifying accidental torsion. Direction of load effects is covered in Section 12.5. For structures in SDC B, the analysis (including accidental torsion effects) may be performed independently in each direction and the structure may be designed on that basis. For structures in higher SDCs, the direction of load effects must be explicitly considered wherever a Type 5 horizontal nonparallel systems irregularity occurs (Table 12.3-1). Because such irregularities are common, many buildings must be designed for a complicated combination of loads that include gravity, lateral loads acting from any direction, simultaneous application of lateral loads acting in orthogonal directions, and accidental torsion with or without amplification.

The manner in which the different load effects are considered depends on whether the analysis is being performed using the equivalent lateral force (ELF) method or the modal response spectrum (MRS) method. Procedures used in association with response history analysis are beyond the scope of this *Guide*.

Before illustrating the procedures used in association with ELF or MRS analysis, it is important to note that these procedures, as presented herein, depend on the analysis being performed in three dimensions. Section 12.7.3 requires 3D modeling for structures with horizontal structural irregularities of Type 1a, 1b, 4, or 5 of Table 12.3-1. Additionally, a 3D analysis is required for structures with semirigid diaphragms. Even where 3D analysis is not required by ASCE 7, using such analysis is advisable because the requirements for accidental torsion and loading direction are easier to apply than would be the case if the structure were to be decomposed into 2D models.

Load Combination Procedures Used in ELF Analysis

In ELF analysis, as many as 16 seismic lateral load cases may be required. The generation of the 16 lateral load cases is shown in **Table G15-1** and **Fig. G15-3**.

The first column of **Table G15-1** represents the direct lateral load without accidental eccentricity. These forces would come from Eqs. (12.8-1), (12.8-11), and (12.8-12). The second column provides the eccentricity at which the lateral loads must be applied. As described in Section 12.8.4.2, this eccentricity is equal to at least 0.05 times the dimension of the building perpendicular to the direction of the applied loads. Both positive and negative eccentricities must be considered for each direction of lateral load. If the building is in SDC C or higher and has a torsional or extreme torsional irregularity, the accidental torsion must be amplified per Section 12.8.4.3. The torsion amplifier, given by Eq. (12.8-14), may be different at each level of the structure.

The third column of **Table G15-1** represents the orthogonal loading requirements, which are specified in Section 12.5. When orthogonal load is required in SDC C and higher, satisfying these requirements by simultaneously applying 100% of the load in one direction (with torsion and torsion amplification if necessary) and 30% of the orthogonal direction loading is permitted. Section 12.5.4 provides additional orthogonal loading requirements for certain interacting structural components in structures that have

Table G15-1   Generation of ELF Load Cases

Major Load Direction	Major Load Applied at Eccentricity[a]	Orthogonal Load (applied at zero eccentricity)[b]	Load Case Number
$+V_{EW}$	$0.05A_xB$	$+0.3\,V_{NS}$	1
		$-0.3\,V_{NS}$	2
	$-0.05A_xB$	$+0.3\,V_{NS}$	3
		$-0.3\,V_{NS}$	4
$-V_{EW}$	$0.05A_xB$	$+0.3\,V_{NS}$	5
		$-0.3\,V_{NS}$	6
	$-0.05A_xB$	$+0.3\,V_{NS}$	7
		$-0.3\,V_{NS}$	8
$+V_{NS}$	$0.05A_xL$	$+0.3\,V_{EW}$	9
		$-0.3\,V_{EW}$	10
	$-0.05A_xL$	$+0.3\,V_{EW}$	11
		$-0.3\,V_{EW}$	12
$-V_{NS}$	$0.05A_xL$	$+0.3\,V_{EW}$	13
		$-0.3\,V_{EW}$	14
	$-0.05A_xL$	$+0.3\,V_{EW}$	15
		$-0.3\,V_{EW}$	16

[a]A_x is the torsional amplification factor.

[b]Not always required. See Section 12.5.

been assigned to SDC D though F. According to Section 12.8.4.2, the orthogonal direction load need not be applied with an eccentricity.

Load Combination Procedures Used in MRS Analysis

Where the modal response spectrum method of analysis is used, all signs in the member forces are lost because of the square root of the sum of the squares (SRSS) or complete quadratic combination (CQC) modal combinations. Additionally, applying accidental torsion as a static load and then combining this load with the results of the modal analysis is common. Orthogonal load effects may be handled in one of two manners:

1. Apply 100% of the spectrum in one direction, and run a separate analysis with 30% of the spectrum in the orthogonal direction. Member forces and displacements are obtained by SRSS or CQC for each analysis. Combine the two sets of results by direct addition.
2. Apply 100% of the spectrum independently in each of two orthogonal directions. Member forces and displacements are found by CQC. Combine the two sets of results by taking the SRSS of results from the two separate analyses.

Fig. G15-3
16 basic lateral load cases used in ELF analysis

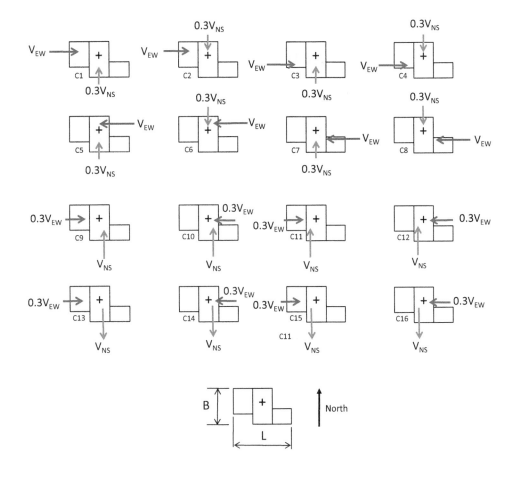

Fig. G15-4
Static load cases for torsion and MRS analysis

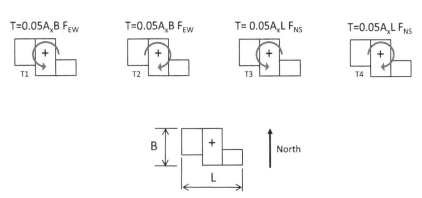

The first method gives different results for different angles of attack for the main component of loading. The main advantage of the second method is that it produces the same results regardless of the angle of attack of the seismic loads (Wilson 2004). From either of these approaches, only two dynamic load analyses are required.

The results from the gravity and response spectrum analysis are then combined algebraically with the results of static accidental torsion analyses, where the accidental torsion, amplified if necessary, is applied. There are only four basic cases of accidental torsion loading. These cases are illustrated in **Fig. G15-4**.

Special Seismic Load Combinations, Including the Overstrength Factor

In some cases, designing members or connections for load effects, including the overstrength factor Ω_o, may be necessary. This factor is listed for each viable system in Table 12.2-1. The requirement to use the overstrength factor may come directly from ASCE 7, or it may come from the specification used to proportion and detail the member or connection. Examples of ASCE 7 requiring the use of the overstrength factor include elements supporting discontinuous walls or frames (Section 12.3.3.3) and the design of diaphragm collector elements (Section 12.10.2.1). An example from a material specification, in this case the *Seismic Provisions for Structural Steel Buildings* (AISC 2010b), is the requirement that moment connections in intermediate moment frames be designed for shear forces that are determined by "using load combinations stipulated by the applicable building code including the amplified seismic load." Similar provisions exist for steel columns and anchor rods. The amplified seismic load is defined as the "horizontal component of earthquake load E multiplied by Ω_o."

The special load combinations that include the overstrength factor are

Combination 5: $(1.2 + 0.2S_{DS})D + \Omega_o Q_E + 1.0L + 0.2S$, and

Combination 7: $(0.9 - 0.2S_{DS})D + \Omega_o Q_E$.

Where the only difference with respect to the standard load combination is that the term Ω_o replaces the redundancy factor ρ. Note that the factor 1.0 on the live load in combination 5 may be reduced to 0.5 when the unreduced live load, L_o, is less than 100 psf. Again, load combinations including the overstrength factor are required only for a few select members or connections. Most members and connections are designed using the standard load combination. In no circumstance are both the redundancy factor and the overstrength factor used at the same time.

Example 16

Effective Seismic Weight (Mass)

In this example, the effective seismic weight is computed for an office and warehouse building in Burlington, Vermont. The example demonstrates the requirements for including both storage live load and snow load in the effective weight calculations. An additional example is provided to illustrate the computation of effective seismic weight for a one-story building with heavy self-supporting wall panels.

Four-Story Book Warehouse and Office Building in Burlington, Vermont

The building in this example is an office and warehouse building in Burlington, Vermont. Plans and an elevation of the building are shown in **Fig. G16-1**. The first floor, at grade level, is used for both storage and office space, with about 70% of the area dedicated to storage. The second and third floors are used for storage only, and the fourth floor consists only of office space. The storage area is used primarily for boxes of textbooks to be used in Burlington area public schools.

The structural system for the building is a prestressed concrete flat slab. This system supports gravity loads and acts as a moment-resistant frame for both wind and seismic forces. For seismic design, the frame is classified as an ordinary moment-resisting frame. The equivalent lateral force (ELF) method is likely to be used for the structural analysis of the system.

The slabs, constructed from lightweight structural concrete with a density of 115 lb/ft^3, have a basic thickness of 9 in. and are thickened to 12 in. at the second and third floor column areas to provide resistance to shear. The slabs at the fourth level (the office floor) and the roof do not have drop

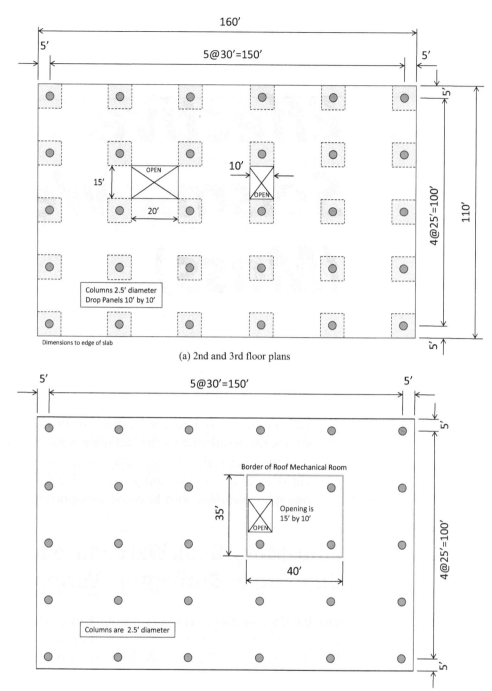

(a) 2nd and 3rd floor plans

(b) Roof plan (4th floor similar, without mechanical room)

panels. A small mechanical penthouse, constructed from steel, is built over the roof slab, as shown in **Figs. G16-1(b)** and **G16-1(c)**. The penthouse is braced laterally and is sufficiently rigid to transfer its tributary roof and snow loads to the main roof level; hence, it is not considered a separate story.

For this example, the columns are also assumed to be constructed from 115 lb/ft³ lightweight concrete. In most cases, normal-weight concrete would be used for the columns of a concrete building.

The building is clad with lightweight precast concrete architectural panels with a thickness of 4 in. The concrete used for these panels has a

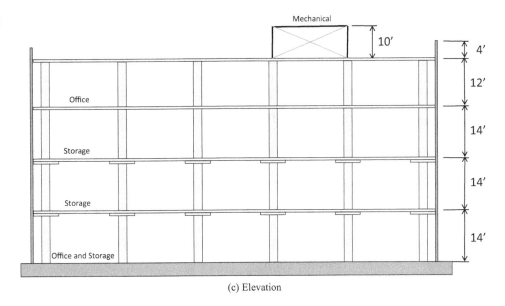

(c) Elevation

density of 90 lb/ft³. These panels have window openings that cover approximately 35% of the façade. These window areas weigh 12 lb/ft². The exterior wall extends 4 ft above the roof to form a solid parapet. The panels are supported vertically at grade and at levels 2, 3, and 4. The detailing of the panel connections is such that the panels are considered as effective seismic weight in each direction.

The books are stored in plastic containers, which in turn are supported by a steel rack system. The racks cover approximately 70% of the floor area. Small forklifts (not to be classified as permanent equipment) are used to place and remove pallets of containers from the shelves. The design live load for the book storage area is 150 lb/ft². The rack storage system, which is anchored to the slab, weighs approximately 20 lb/ft². The racking system is laterally braced in two orthogonal directions with steel X-bracing. The system is sufficiently rigid to transfer the storage loads to the floor slabs.

The office area on the fourth floor of the building is designed for a live load of 50 lb/ft². Various work spaces are formed by a combination of fixed and movable partitions. A partition allowance of 15 lb/ft² is used in the design of the office floors. Two-thirds of this value, 10 lb/ft², is used for effective seismic weight as allowed by item 2 in Section 12.7.2.

The design dead load value used for the ceiling and mechanical areas of the main building is 15 lb/ft². Floor finishes in the office area are assumed to weigh 2.5 lb/ft². The floors in the storage areas are bare concrete.

The second and third floors have two openings, one (15 ft × 20 ft) to accommodate a hydraulic elevator for use in transporting the books (including the small forklifts), and the other (15 ft × 10 ft) for an elevator that services the offices on the fourth floor. The fourth floor and roof have only the smaller opening. Other minor openings exist in the floors and roof, but these openings are small and are not considered when computing the effective seismic weight. Two stairwells are also present in the building (not shown on plan), but the weights of these are, on a pound per square foot basis, approximately the same as the floor slab.

The mechanical room contains various heating, air conditioning, and ventilating equipment. The average dead load for the entire mechanical room, including the steel framing, roof, and equipment, is 60 lb/ft². The roofing over the remainder of the building (that area not covered by the mechanical room) is assumed to weigh 15 lb/ft².

The ground snow load for the building site is 60 lb/ft². Based on the procedures outlined in Chapter 7 of ASCE 7, the flat roof snow load is 42 lb/ft².

Calculation of the effective seismic weight, W, is based on the requirements of Section 12.7.2. The weight includes all dead load, a minimum of 25% of the floor live load in the storage areas, a 10 lb/ft² partition allowance where appropriate, total operating weight of permanent equipment, and 20% of the uniform design snow load when the flat roof snow load exceeds 20 lb/ft². Each of these load types is pertinent to the building under consideration.

Dead Load

The seismic load for the first floor level (a slab on grade) transfers directly into the foundation, so this load need not be considered as part of the effective seismic weight.

The loading for the second floor consists of the slab, drop panels, columns, storage rack system, ceiling and mechanical system, and exterior cladding.

Slab:	Total area = 160 × 110 – 15 × 20 – 15 × 10 = 17,150 ft²,
	Unit weight = (9/12) × 115 = 86.2 lb/ft², and
	Weight = 17,150 × 86.2/1,000 = 1,478 kip.
Drop panels:	30 panels × 100 ft² per panel = 3,000 ft²,
	Unit weight = (3/12) × 115 = 28.8 lb/ft², and
	Weight = 3,000 × 28.8/1,000 = 86 kip.
Columns:	Clear height tributary to first story = 13 ft,
	Clear height tributary to second story = 13 ft,
	Height tributary to second level = (13+13)/2 = 13 ft,
	Column area = 4.91 ft², and
	Weight = 30 columns × 4.91 × 13 × 115/1,000 = 220 kip.
Storage rack system:	Total area = 17,150 ft² (no deduction taken for columns),
	Effective area = 0.7 × 17,150 = 12,005 ft²,
	Unit weight = 20 lb/ft², and
	Weight = 12,005 × 20/1,000 = 240 kips.
Ceiling and mechanical system:	Total area = 17,150 ft² (no deduction taken for columns),
	Unit weight = 15 lb/ft², and
	Weight = 17,150 × 15/1,000 = 257 kip.
Exterior cladding:	Perimeter = 2(160 + 110) = 540 ft,
	Height tributary to second level = 14 ft,

Area of 4-in.-thick precast = $0.65 \times 540 \times 14$ = 4,914 ft^2,

Unit weight of panel = $(4/12) \times 90 = 30.0$ lb/ft^2,

Total panel weight = $4,914 \times 30.0/1,000 = 147$ kip,

Area of glass windows = $0.35 \times 540 \times 14$ = 2,646 ft^2,

Unit weight of glass = 12 lb/ft^2,

Total glass weight = $2,646 \times 12/1,000 = 32$ kip, and

Total cladding weight = $147 + 32 = 179$ kip.

Total dead load at second level = $1,478 + 86 + 220 + 240 + 257 + 179 = 2,460$ kip.

The dead load on the third level is almost identical to that on the second level. The only difference is that the absence of the drop panels at the fourth story has a slight influence on the clear length of the columns at the third story. For this example, this small difference is ignored, and the same dead load is used for the second and third levels.

The dead load for the fourth level is computed as follows:

Slab:	Total area = $160 \times 110 - 15 \times 10 = 17,450$ ft^2,
	Unit weight = $(9/12) \times 115 = 86.2$ lb/ft^2, and
	Weight = $17,450 \times 86.2/1,000 = 1,504$ kip.
Columns:	Clear height tributary to third story = 13.25 ft,
	Clear height tributary to fourth story = 11.25 ft,
	Height tributary to second level = $(13.25 + 11.25)/2 = 12.25$ ft,
	Column area = 4.91 ft^2, and
	Weight = 30 columns $\times 4.91 \times 12.25 \times 115/1,000 = 208$ kip.
Partitions:	Total area = 17,450 ft^2 (no deduction taken for columns),
	Unit weight = 10 lb/ft^2 (see Section 12.7.2, item 2), and
	Weight = $17,450 \times 10/1,000 = 175$ kip.
Floor finish:	Total area = 17,450 ft^2 (no deduction taken for columns),
	Unit weight = 2.5 lb/ft^2, and
	Weight = $17,450 \times 2.5/1,000 = 44$ kip.
Ceiling and mechanical system:	Total area = 17,450 ft^2 (no deduction taken for columns),
	Unit weight = 15 lb/ft^2, and
	Weight = $17,450 \times 15/1,000 = 262$ kip.
Cladding:	Perimeter = $2(160 + 110) = 540$ ft,
	Height tributary to fourth level = $0.5(14 + 12) = 13$ ft,
	Area of 4-in.-thick precast = $0.65 \times 540 \times 13$ = 4,563 ft^2,

Unit weight of panel = (4/12) × 90 = 30 lb/ft²,

Total panel weight = 4,563 × 30/1,000 = 137 kip,

Area of glass windows = 0.35 × 540 × 13 = 2,457 ft²,

Unit weight of glass = 12 lb/ft²,

Total glass weight = 2,457 × 12/1,000 = 29 kip, and

Total cladding weight = 137 + 29 = 166 kip.

Total dead load at fourth level = 1,504 + 208 + 175 + 44 + 262 + 166 = 2,359 kip.

The dead load for the roof level is computed as follows:

Slab:
Total area = 160 × 110 – 15 × 10 = 17,450 ft²,

Unit weight = (9/12) × 115 = 86.2 ft², and

Weight = 17,450 × 86.2/1,000 = 1,504 kip.

Columns:
Clear height at fourth story = 11.25 ft,

Height tributary to second level = (11.25)/2 = 5.62 ft,

Column area = 4.91 ft², and

Weight = 30 columns × 4.91 × 5.62 × 115/1,000 = 95 kip.

Ceiling and mechanical system:
Total area = 17,450 ft² (no deduction taken for columns),

Unit weight = 15 lb/ft², and

Weight = 17,450 × 15/1,000 = 262 kip.

Roofing:
Total area of main roof = 160 × 110 – 40 × 35 = 16,200 ft²,

Unit weight = 15 lb/ft², and

Weight = 16,200 × 15/1,000 = 243 kip.

Mechanical area:
Total area = 40 × 35 = 1,400 ft²,

Unit weight = 60 lb/ft² (estimated), and

Weight = 1,400 × 60/1,000 = 84 kip.

Cladding:
Perimeter = 2(160 + 110) = 540 ft,

Height tributary to roof = 6 ft,

Area of 4-in.-thick precast = 0.65 × 540 × 6 = 2,106 ft²,

Unit weight of precast = (4/12) × 90 = 30 lb/ft²,

Total panel weight = 2,106 × 30/1,000 = 63 kip,

Area of glass windows = 0.35 × 540 × 6 = 1,134 ft²,

Unit weight of glass = 12 lb/ft²,

Total glass weight = 1,134 × 12/1,000 = 14 kip, and

Total cladding weight = 63 + 14 = 77 kip.

Parapet:
Perimeter = 2(160 + 110) = 540 ft,

Height tributary to roof = 4 ft,

Area of 4-in.-thick precast = 540 × 4 = 2,160 ft²,

Unit weight of precast = $(4/12) \times 90 = 30$ lb/ft^2, and

Total parapet weight = $2{,}160 \times 30/1{,}000 = 65$ kip.

Total dead load at the roof = $1{,}504 + 95 + 262 + 243 + 84 + 77 + 65 = 2{,}330$ kip.

The total dead weight for the building, including the mechanical level, is 9,609 kip. Using a building volume exclusive of the mechanical area of $(160 \times 110) \times 54 = 950{,}400$ ft^3, the dead load density for the building is $9{,}609/950{,}400 = 0.0101$ kip/ft^3 or 10.0 lb/ft^3. This weight is a bit heavier than that which would be appropriate for a low-rise office building, but it is reasonable for a concrete warehouse building. Calculation of building density is a good reality check on effective seismic weight. Low-rise buildings generally have a density in the range of 7 to 10 lb/ft^3, depending on material and use.

Contribution from Storage Live Loads at Levels 2 and 3

As mentioned in the building description, the building has a design storage live load of 150 lb/ft^2. However, only 70% of each floor is reserved for storage, and the remainder is used for aisles, stairs, and restrooms. The openings for elevators are considered separately.

Building use statistics indicate that the storage racks are near capacity in the summer months when school is not in session and reduce to about 30% capacity during the fall and winter months. Section 12.7.2 states that a minimum of 25% of storage live load shall be used as effective seismic weight. For this facility, the 25% minimum is used. However, others might argue that, on the basis of use statistics, a larger portion of the load should be used.[1]

The live load contribution to effective seismic weight is as follows for the second and third levels:

Total area = $160 \times 110 - 15 \times 20 - 15 \times 10 = 17{,}150$ ft^2,
Effective area = $0.7 \times 17{,}150 = 12{,}005$ ft^2,
Effective live load = $0.25(150) = 37.5$ lb/ft^2, and
Total live load contribution to seismic weight = $12{,}005 \times 37.5/1{,}000 = 450$ kip.

Contribution of Snow Load at Roof Level

Section 12.7.2 indicates that 20% of the uniform design snow load must be included in the effective seismic weight when the flat roof snow load exceeds 30 lb/ft^2. The flat roof snow load for this building is 42 lb/ft^2, so snow load

---

[1] The design of combined building-rack storage systems is considerably more complex than indicated in this example. See Chapters 13 and 15 of ASCE 7 for requirements for the design and attachment of rack systems to the building superstructure. See also the *Specifications for the Design, Testing, and Utilization of Industrial Steel Storage Racks* (RMI 2009).

Table G16-1   Summary of Effective Seismic Weight Calculations

| Level | Load Contribution (kip) | | | |
	Dead	Live	Snow	Total
2	2,460	450	0	2,910
3	2,460	450	0	2,910
4	2,359	0	0	2,359
R	2,330	0	148	2,478
Total	9,609	900	148	10,657

must be included. The building has a flat roof (both the main roof and the mechanical room), so the uniform snow load is 42 lb/ft². Using a total area of $160 \times 110 = 17,600$ ft², the contribution from snow to the effective seismic weight is

Total area = 17,600 ft²,
Effective snow load = 0.2(42) = 8.4 lb/ft², and
Total snow load contribution to seismic weight = $17,600 \times 8.4/1,000$
= 148 kip.

Table G16-1 summarizes the effective seismic weight for the entire system. The design seismic base shear [Eq. (12.8-1)] should be based on these weights, as should the distribution of forces along the height of the building [Eqs. (12.8-11) and (12.8-12)]. These forces should be placed at the center of mass of floors of the building, as appropriate. For this building, the center of mass is slightly offset from the plan center because of the floor openings and the somewhat eccentric location of the mechanical room.

The weights shown in **Table G16-1** are to be used in an ELF analysis of the system. If a three-dimensional rigid-diaphram modal analysis is used, the mass moments of inertia are required for each floor. When heavy cladding is used, including this cladding as line masses situated at the perimeter may be appropriate.

If the mechanical penthouse covered more area, considering this as a separate level of the building may be appropriate. Section 12.2.3.1 covers situations when different lateral load–resisting systems are used along the height of the building. The light rooftop structure used in this Seismic Design Category B example is exempt from the requirements of 12.2.3.1.

Consideration should also be given to the design, detailing, and anchorage of the steel rack system used in this building. Chapters 13 and 15 provide the requirements for the analysis and design of the system.

Low-Rise Industrial Building

In the previous example, the weight of the cladding parallel and perpendicular to the direction of loading was included in the effective seismic weight in each direction. Thus, the effective seismic weight is the same in each direction.

Fig. G16-2

Plan view of a one-story industrial building

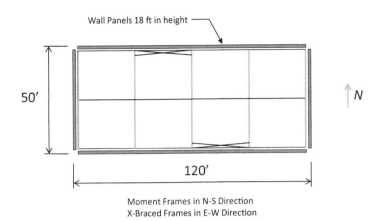

Wall Panels 18 ft in height

50'

120'

N

Moment Frames in N-S Direction
X-Braced Frames in E-W Direction

For low-rise buildings, typically one story, the cladding panels may be detailed such that they are self-supporting when resisting seismic loads parallel to the plane of the panels and hence do not contribute to the seismic resistance of the main structural system when seismic loads act parallel to the plane of the panels. However, the effective weight of panels perpendicular to the direction of loading must be included.

Consider the low-rise industrial building shown in **Fig. G16-2**. The siding for the building consists of 5-in.-thick insulating concrete sandwich panels that weigh 60 lb/ft^2. On the west face, openings cover approximately 35% of the wall panel area. The other faces have only minor window and door openings, and these openings are ignored in computing the effective seismic weight. Only the panel weight is considered herein.

For seismic forces in the north–south direction, the panel contribution to the effective seismic weight is based on one-half of the weight of the two 120-ft-long walls:

$$W_{panels,N-S} = 2 \times 120 \text{ ft} \times 18 \text{ ft} \times 60 \text{ lb/ft}^2 \times 0.5 = 129,600 \text{ lb} = 130 \text{ kip.}$$

For seismic loads in the east–west direction, the effective seismic weight is

$$W_{panels,E-W} = (1 + 0.65) \times 50 \text{ ft} \times 18 \text{ ft} \times 60 \text{ lb/ft}^2 \times 0.5 = 44,550 \text{ lb} = 45 \text{ kip.}$$

The 0.5 in the aforementioned calculations is based on half of the total effective seismic panel weight being carried by the steel roof framing, and the remainder being resisted at the foundation.

Example 17

Period of Vibration

This example explores computing the period of vibration of building structures. This example reviews the empirical methods that ASCE 7 provides for computing periods, computes periods for a few simple buildings, and provides a more detailed analysis wherein the empirical periods are compared with the period based on rational analysis.

Approximate Fundamental Period T_a

Section 12.8.2.1 addresses computing the approximate period of vibration of buildings. Three basic formulas are provided:

$$T_a = C_t h_n^x \qquad \text{(Eq. 12.8-7)}$$

$$T_a = 0.1\,N \qquad \text{(Eq. 12.8-8)}$$

$$T_a = \frac{0.0019}{\sqrt{C_w}}\, h_n \qquad \text{(Eq. 12.8-9)}$$

These formulas are highly empirical and are to be used for seismic analysis of building structures only. Eq. (12.8-7) applies to all buildings, Eq. (12.8-8) applies to certain moment frames, and Eq. (12.8-9) applies only for masonry or concrete shear wall structures. The primary use of T_a is in computing seismic base shear V. The period (in relation to T_S) is also used to determine the appropriate method of analysis (Table 12.6-1).

In Eq. (12.8-7), the coefficient C_t and the exponent x come from Table 12.8-2 and depend on the structural system and structural material. These terms were developed from regression analysis of the measured periods of real buildings in California. The coefficients for buckling-restrained brace (BRB) systems are the same as those for eccentrically braced frames ($C_t = 0.03$, $x = 0.75$). These coefficients produce a somewhat longer period than those obtained for braced frames (all other systems in the table). This is logical because the cross-sectional area of the core bracing elements in BRB systems is always smaller than that required for traditional braced frames. Note also that the coefficients used for eccentrically braced steel frames apply only if the eccentrically braced frame is designed and detailed according to the requirements of the *Seismic Provisions for Structural Steel Buildings* (AISC 2010b).

When applying Eq. (12.8-7), the basic uncertainty is in the appropriate value to use for the structural height h_n, which is defined as "the vertical distance from the base to the highest level of the seismic force resisting system of the structure." Section 11.2 of ASCE 7 defines the base as "the level at which the horizontal seismic ground motions are considered to be imparted to the structure." For a building on level ground without basements, the base may be taken as the grade level. In many cases, however, establishing the exact location of the base may not be easy. This is particularly true when the building is constructed on a sloped site, or when one or more basement levels exist. Also of some concern is the definition of the highest level of the structure. This height should not include small mechanical rooms or other minor rooftop appurtenances. For buildings with sloped roofs, the structural height should be taken from the base to the average height of the roof. See also Commentary Section 12.2 for definitions and illustrations of the base.

Consider, for example, the X-braced steel frame structures shown in **Fig. G17-1**. Structure (a) has no basement. Here, the height h_n is the distance from the grade level to the roof, not including the penthouse. Structure (b) is like structure (a), but has a full basement. At the grade level, the slab is thickened, and horizontal seismic force at the grade level is partially transferred though the diaphragms to exterior basement walls. Here again, the effective height should be taken as the distance from the grade level to the main roof. However, a computer model would produce a longer period for structure (b) compared with structure (a) because of the axial deformations that occur in the subgrade braces and columns of the braced frame of structure (b). In structure (c), the first grade slab is not thickened, and the braced frame shear forces are not expected to completely transfer out through the first-floor diaphragm. However, the soils adjacent to the basement walls

Fig. G17-1

Finding the effective height h_n for a braced frame

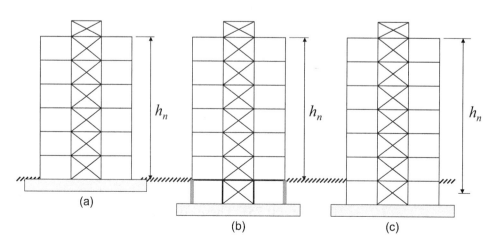

provide some lateral support. Here, the period of vibration is longer than for either (a) or (b), but may not be long as determined using the distance from the main roof to the top of the basement slab. Thus, a height from the main roof to the midlevel of the basement might be appropriate. Engineering judgment may be required. When in doubt, use the shortest reasonable height for h_n, because this height produces the most conservative base shear.

Using structure (a) as an example, assume that the story height is 13 ft. Hence, $h_n = 13 \times 6 = 78$ ft. From Table 12.8-2, the "All other structural systems" classification applies, giving $C_t = 0.02$ and $x = 0.75$. Using Eq. (12.8-7),

$$T_a = C_t h_n^x = 0.02(78)^{0.75} = 0.525 \text{ s for the steel braced frame.}$$

If a steel moment frame is used in lieu of the braced frame, Table 12.8-2 gives $C_t = 0.028$ and $x = 0.80$. Using Eq. (12.8-7),

$$T_a = C_t h_n^x = 0.028(78)^{0.8} = 0.914 \text{ s for the steel moment frame.}$$

The longer period in this case reflects the fact that moment frames are generally more flexible than braced frames. Eq. (12.8-8) is applicable for the moment frame and produces

$$T_a = 0.1 \, N = 0.1(6) = 0.60 \text{ s for the steel moment frame.}$$

This shorter period is somewhat more conservative, which leads to a larger base shear than that given by Eq. (12.8-7).

Period T Used in the Equivalent Lateral Force Method

The total base shear and the distribution of lateral forces along the height of the structure are both functions of the period of vibration, T. To compute these quantities, the approximate period T_a may be used, giving $T = T_a$. This choice is generally conservative because periods computed on the basis of a rational structural analysis are almost always greater than those computed from the empirical formulas. In recognition of this fact, ASCE 7 (Section 12.8.2) allows the use of a modified period $T = C_u T_a$, where the coefficient C_u is provided in Table 12.8-1. The modifier may be used only if a period, called $T_{computed}$ herein, is available from a properly substantiated structural analysis.

Consider the braced frame structure [**Fig. G17-1(a)**] discussed earlier, with $T_a = 0.525$ s. A computer analysis has predicted a period of $T_{computed} = 0.921$ s for the structure, so the modified period $T = C_u T_a$ may be used. Assuming that the structure is in a region of moderate seismicity with $S_{D1} = 0.25 \, g$ and interpolating from Table 12.8-1,

$$C_u = 0.5(1.4 + 1.5) = 1.45, \text{ and}$$

$$T = C_u T_a = 1.45 \times 0.525 = 0.761 \text{ s.}$$

The period of 0.761 s must be used to determine the set of equivalent lateral forces from which the strength of the structure will be evaluated, even though the rational period $T_{computed}$ was somewhat longer at 0.921 s. As explained in a separate example (Example 19), Section 12.8.6.2 of the standard allows the computed period to be used in the development of an alternate set of equivalent lateral forces that are used only for drift calculations.

When Computed Period T$_{computed}$ Is Less than T = C$_u$T$_a$

The computed period may be less than the upper limit period $C_u T_a$. Continuing with the braced frame, assume that the computed period $T_{computed} = 0.615 < C_u T_a = 0.761$ s. Although ASCE 7 is silent on this possibility, it is recommended that the lower period be used in the calculations. In the unlikely event that the computed period turns out to be less than $T_a = 0.525$ s, the period $T = T_a = 0.525$ s may be used because there is no requirement that $T_{computed}$ shall be determined. If the computed period is significantly different from $C_u T_a$, say less than 0.5 $C_u T_a$ or more than 2 $C_u T_a$, the computer model should be carefully inspected for errors.

Table G17-1 summarizes the period values that should be used in the strength and drift calculations.

Period Computed Using Equivalent Lateral Forces and Resulting Displacements

Most commercial structural analysis computer programs can calculate the periods of vibration of building structures. If this capability is not available, the period can be estimated from the displacements produced from a set of lateral forces. The lateral force distributions provided by Eqs. (12.8-11) and (12.8-12) are well suited to this calculation, but these forces depend on the exponent k, which in turn depends on T, which is the goal of the calculation. For the purpose of computing the lateral forces required for the period calculation, it is recommended that k be based on a trial period of $T = C_u T_a$.

The formula for computing the approximate period is based on a first-order Rayleigh analysis and is as follows:

$$T = 2\pi \sqrt{\frac{\sum_{i=1}^{n} \delta_i^2 W_i}{g \sum_{i=1}^{n} \delta_i F_i}}$$

(Eq. G17-1)

Table G17-1   Summary of Period Values to Be Used in Calculations

Situation	Period T to Be Used in Strength Calculations	Period T to Be Used in Drift Calculations
$T_{computed} \leq T_a$	T_a	T_a
$T_a < T_{computed} < C_u T_a$	$T_{computed}$	$T_{computed}$
$T_{computed} \geq C_u T_a$	$C_u T_a$	$T_{computed}$

Table G17-2   Determination of Period Using the Analytical Method

Level i	F_i (kip)	δ_i (in.)	W_i (kip)	$\delta_i F_i$ (kip-in.)	$\Sigma\delta_i^2 W_i$ (kip-in.^{2})
6	301	3.95	1,100	1,189	17,163
5	245	3.15	1,100	772	10,914
4	190	2.40	1,100	456	6,336
3	137	1.70	1,100	233	3,179
2	87	1.05	1,100	91	1,213
1	40	0.5	1,100	20	275
				$\Sigma = 2{,}761$	$\Sigma = 39{,}080$

where F_i is the lateral force at level i, δ_i is the lateral displacement at level i, W_i is the weight at level i, n is the number of levels, and g is the acceleration of gravity. When using the equation, the displacements δ are those that result directly from the application of the lateral forces F, and they do not include the deflection amplifier C_d.

The procedure is illustrated in **Table G17-2** for the braced frame in **Fig. G17-1(a)**. For the example, the story weights are assumed to be uniform at 1,100 kip per level. The mechanical penthouse is not included in the analysis. Using $T = C_u T_a = 0.761$ s, $k = 1.13$. The period computed from the information provided in **Table G17-2** is as follows:

$$T = 2\pi\sqrt{\frac{1}{386.4}\frac{39080}{2761}} = 1.20 \text{ s}$$

Period Computed Using Computer Programs

Most commercially available structural analysis programs can calculate the period of vibration. The computed period depends directly on the assumptions made in modeling the mass and stiffness of the various components of the structure and on the manner in which the boundary conditions (supports) are represented. In many cases, determining the appropriate component stiffness is not straightforward because these properties depend on several factors, including the effective rigidity of connections, the degree of composite action, and the degree of cracking in concrete. Although providing detailed information on modeling is beyond the scope of this *Guide*, the following points are noted.

Deformations in the panel zones of the beam–column joints of steel moment frames are a significant source of flexibility in these frames. Two mechanical models for including such deformations are summarized in Charney and Marshall (2006). These methods are applicable to both elastic and inelastic systems. For elastic structures, the use of centerline analysis provides reasonable, but not always conservative, estimates of frame flexibility. Fully rigid end zones, as allowed by many computer programs, should never be used because they always result in an overestimation of lateral stiffness in steel moment-resisting frames. The use of partially rigid end zones may be justified in certain cases, such as when doubler plates are used to reinforce the panel zone. Partially rigid end zones are also appropriate in the modeling of the joints of reinforced concrete buildings.

The effect of composite slabs on the stiffness of beams and girders may be warranted in some circumstances. When composite behavior is included, due consideration should be paid to the reduction in effective composite stiffness when portions of the slab are in tension (Schaffhausen and Wegmuller, 1977; Liew, 2001).

For reinforced concrete buildings, representing the effects of axial, flexural, and shear cracking in all structural components is necessary. Recommendations for computing cracked section properties may be found in Paulay and Priestly (1992) and other similar texts. In terms of the degree of cracking to use in the analysis, Table 6-5 of ASCE 41-06 (ASCE, 2007) provides reduction factors that can be used to represent flexural cracking. For example, ASCE 41-06 recommends that 50% percent of the gross section moment of inertia be used in modeling beams and girders in non-prestressed structures and that 70% of the gross section moment of inertia be used for columns in compression. ASCE 41-06 does not provide reduction factors for axial or shear rigidity due to cracking. In the author's opinion this is not appropriate as the reduction in shear stiffness due to cracking can be significant (Park and Paulay, 1974).

Shear deformations should be included in all structural analyses. Such deformations can be significant in steel moment frames. Additionally, shear deformations in reinforced concrete shear wall systems can actually dominate the flexibility when the walls have height-to-width ratios of less than 1.0. Finally, it is important to include both flexural and shear cracking in the coupling beams of coupled wall systems.

Computing T_a for Masonry and Concrete Shear Wall Structures

Section 12.8.2.1 prescribes an alternate method for computing T_a for masonry or concrete shear wall structures. Two equations are provided:

$$T_a = \frac{0.0019}{\sqrt{C_W}} h_n \qquad \text{(Eq. 12.8-9)}$$

$$C_W = \frac{100}{A_B} \sum_{i=1}^{x} \frac{A_i}{\left[1 + 0.83\left(\dfrac{h_i}{D_i}\right)^2\right]} \qquad \text{(Eq. 12.8-10)}$$

where

A_B = area of the base of the structure (ft^2),
A_i = area of the web of wall i (ft^2),

D_i = plan length of wall i (ft), and

x = number of walls in the direction under consideration.

The use of the equations is based on a paper by Goel and Chopra (1997) and is limited to buildings with a structural height less than or equal to 120 ft and for which all walls extend the full height of the building. These equations are exercised using the structure shown in **Fig. G17-2**. The building has eight walls, with dimensions shown in the figure. The analysis is performed with four different assumptions on the height of the building. In all cases, the total building area is 15,000 ft^2.

The results of the analysis are shown in **Table G17-3** together with the period computed using Eq. (12.8-7), using $C_t = 0.02$ and $x = 0.75$. As seen in the table, the 30-ft tall system has a period of 0.146 s when Eqs. (12.8-9) and (12.8-10) are used and increases to 0.256 s when Eq. (12.8-7) is utilized. For the other three system heights the period computed using Eqs. (12.8-9) and (12.8-10) is greater than that determined from Eq. (12.8-7). For the 120 ft tall system, the period computed using Eqs. (12.8-9) and (12.8-10) is 2.07 s, which is more than twice the period computed using Eq. (12.8-7).

Given the very large discrepancy between the computed periods using the different formulas for T_a, the engineer should use caution when using Eqs. (12.8-9) and (12.9-10), particularly for structures taller than 60 ft. in height. Additionally, it is the author's opinion that the use of the formula

Fig. G17-2

Concrete shear wall system used for period determination

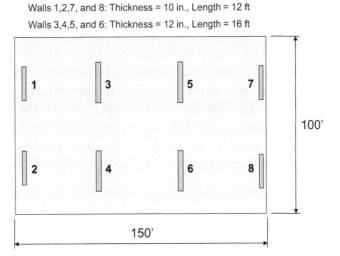

Walls 1,2,7, and 8: Thickness = 10 in., Length = 12 ft

Walls 3,4,5, and 6: Thickness = 12 in., Length = 16 ft

Table G17-3   Computing the Period for Masonry and Concrete Shear Wall Structures

Building Height (ft)	Using Eqs. (12.8-9) and (12.8-10)		Using Eq. (12.8-7) with $C_t = 0.02$ and $x = 0.75$
h_n	C_W	T_a (s)	T_a (s)
30	0.1520	0.146	0.256
60	0.0459	0.532	0.431
90	0.0212	1.173	0.584
120	0.0121	2.070	0.725

$T = C_u T_a$ is not applicable when Eqs. (12.8-9) and (12.8-10) are utilized [or that C_u should be taken as 1.0 when Eqs. (12.8-9) and (12.8-10) are used].

A better approach for determining the period of vibration of shearwall system is to model the system accurately in a finite element analysis program. This is done quite conveniently with modern software.

Periods of Vibration for Three-Dimensional Systems

A computer program is required to determine the periods of vibration for 3D structures. The program reports periods (or frequencies) for as many modes as the user requests. For regular rectangular buildings (regular in the sense of structural irregularities as described in Section 12.3.2), the first three modes generally represent the two orthogonal lateral modes and the torsional mode. These modes can occur in any order, and it is possible (although not generally desirable) that the first or second mode can be a torsional mode. The first mode (or any of the first several modes) can also represent the vibration of a flexible portion of the system, such as a long cantilever, a long span beam, or even an error in connectivity. Thus, before using the periods in an analysis, the engineer should plot and animate the mode shapes to make sure that the proper periods are being used. Reviewing the mass participation of each mode is also a good indicator of structural behavior.

Interpretation and use of the modes and the periods of irregular buildings is often complex because even the first mode shape may represent a coupled lateral–torsional response. The period associated with the mode with the highest mass participation factor in the direction under consideration should be used to determine the seismic forces in that direction. As with regular systems, the mode shapes should always be plotted and animated to make sure that they are reasonable. See Example 20 of this guide for more discussion on mode shapes and periods of vibration in highly irregular buildings.

Example 18

Equivalent Lateral Force Analysis

In this example, an equivalent lateral force (ELF) analysis is performed for a reinforced concrete structure with intermediate moment frames resisting load in one direction and shear walls resisting the load in the other direction. Several aspects of the analysis are considered, including the determination of the design base shear, distribution of lateral forces along the height, application of accidental torsion, and orthogonal load effects. Also included in a separate example is a two-stage ELF procedure (Section 12.2.3.2), which may be applicable when the lower portion of the structure is significantly stiffer than the upper portion.

Before beginning the examples a review of the pertinent equations used for ELF analysis is useful. The first equation provides the design base shear in the direction under consideration:

$$V = C_s W \qquad \text{(Eq. 12.8-1)}$$

where W is the effective seismic weight of the building, and C_s is the seismic response coefficient. W is determined in accordance with Section 12.7.2, and C_s is calculated using one of five equations, numbered (12.8-2) through (12.8-6). The first three of these equations define an inelastic design response spectrum, and the last two provide lower limits on C_s.

The first three equations are as follows:

$$C_s = \frac{S_{DS}}{\left(\dfrac{R}{I_e}\right)} \text{ (applicable when } T \le T_S) \qquad \text{(Eq. 12.8-2)}$$

$$C_S = \frac{S_{D1}}{T\left(\dfrac{R}{I_e}\right)} \quad \text{(applicable when } T_s < T \leq T_L) \tag{Eq. 12.8-3}$$

$$C_S = \frac{S_{D1}T_L}{T^2\left(\dfrac{R}{I_e}\right)} \quad \text{(applicable when } T > T_L) \tag{Eq. 12.8-4}$$

The engineering units for C_s are in terms of the acceleration of gravity, g. This fact is not immediately clear from Eqs. (12.8-3) and (12.8-4) because both equations appear to have units of gravity per second. The apparent inconsistency in the units occurs because of the empirical nature of the equations. Eqs. (12.8-2) through (12.8-4) are plotted as an acceleration spectrum in **Figs. G18-1**. The specific values used to generate the figure are $S_S = 1.0$, $S_1 = 0.33$, Site Class B, $F_a = 1.0$, $F_v = 1.0$, $S_{DS} = 0.667$, $S_{D1} = 0.222$, $T_L = 4.0$, $I_e = 1$, and $R = 6$.

Two transitional periods separate the three branches of the response spectrum. These periods are determined as follows:

$$T_S = \frac{S_{D1}}{S_{DS}} \tag{Section 11.4.5}$$

$$T_L \tag{See Figs. 22-12 to 22-16}$$

T_s was derived by equating Eqs. (12.8-2) and (12.8-3) to find the period at which both branches of the spectrum produce the same value of C_s. Using the relationships provided in Chapter 11, it may be shown that T_s is the product of two ratios:

$$T_s = \frac{S_1}{S_S} \times \frac{F_v}{F_a} \tag{Eq. G18-1}$$

Fig. G18-1
Inelastic design
response spectrum

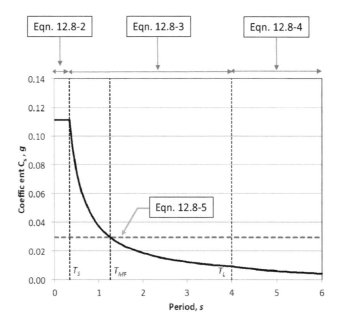

The ratio of S_1 to S_S varies across the United States, ranging from a low of about 0.2 (in New York City, for example), to a high of about 0.4 (in San Jose, for example). The ratio of F_v to F_a is a function of site class and seismicity. For Site Classes A and B, the ratio is fixed at 1.0. For higher site classes, the ratio increases with ground motion intensity and is a maximum of $2.4/0.9 = 2.67$ for Site Class E with $S_S \geq 1.25$ and $S_1 \geq 0.5$. Using the product of the ratios, the transition period T_S can range from a low of about 0.2 s to a high of about 1.07 s.

The transitional period T_L comes from maps provided in Figs. 22-12 through 22-16. The minimum value of T_L is 4.0 s, and this value is applicable only in the Rocky Mountain region and the northwestern Hawaiian islands. Elsewhere, T_L ranges from 6.0 to 16.0 s.

Two equations are used to determine the minimum value of C_s:

$$C_S = 0.044\,S_{DS}I_e \geq 0.01 \text{ (applicable when } S_1 < 0.6\ g) \qquad \text{(Eq. 12.8-5)}$$

$$C_S \geq \frac{0.5S_1}{\left(\dfrac{R}{I_e}\right)} \text{ (applicable when } S_1 \geq 0.6\ g) \qquad \text{(Eq. 12.8-6)}$$

Eq. (12.8-6) controls only in areas of high seismicity and is intended to account for the effects of near-field earthquakes. Eqs. (12.8-3) and (12.8-6), when set equal to each other, produce the period at which Eq. (12.8-6) controls C_s. This period, called T_{MN} herein, is

$$T_{MN} = 1.33F_v \qquad \text{(Eq. G18-2)}$$

As seen in Table 11.4-2, when $S_1 > 0.5\,g$, F_v ranges from 0.8 for Site Class A to 2.4 for Site Class E, thus T_{MN} ranges from 1.06 s to 3.19 s. Clearly, Eq. (12.8-4) is not needed when Eq. (12.8-6) is applicable.

The period at which Eq. (12.8-5) controls may be determined by setting Eqs. (12.8-3) and (12.8-5) equal. This period, called T_{MF} herein, is

$$T_{MF} = \frac{22.7T_S}{R} \qquad \text{(Eq. G18-3)}$$

This transitional period and a horizontal line representing Eq. (12.8-5) are shown in **Fig. G18-1**. The calculations are as follows:

$$T_S = S_{D1}/S_{DS} = 0.222/0.667 = 0.33\text{ s}$$
$$T_{MF} = 22.7T_S/R = 22.7 \times 0.33/6 = 1.26\text{ s}$$
$$C_s \text{ (minimum)} = 0.044S_{DS}I_e = 0.044 \times 0.667 \times 1.0 = 0.029\,g$$

Given the large range of values for T_S and R, developing a feel for T_{MF} is difficult. To help alleviate this problem, **Fig. G18-2** plots the period values, T_{MF} (vertical axis) versus the short period spectral acceleration S_S (horizontal axis) at which Eq. (12.8-5) controls. The plots were developed for the case where S_1/S_S is 0.30. Somewhat different values can be obtained for other reasonable values of S_1/S_S. One graph is provided for each R value in the

Fig. G18-2

Period at which Eq. (12.8-5) controls C_s value (developed using $S_1/S_S = 0.3$)

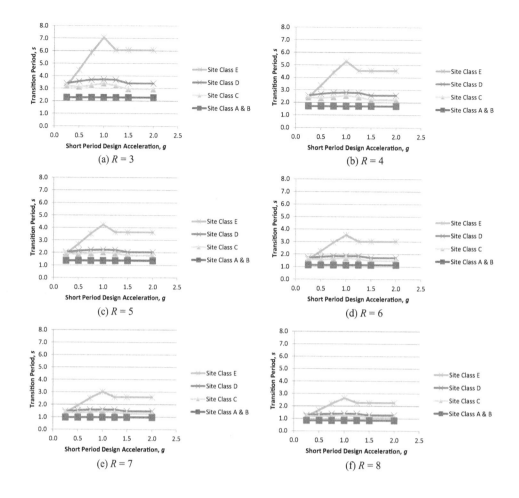

(a) $R = 3$

(b) $R = 4$

(c) $R = 5$

(d) $R = 6$

(e) $R = 7$

(f) $R = 8$

range 3 to 8, inclusive. For lower R values and firmer soils, Eq. (12.8-5) does not control until the period is greater than about 2.0 s. For low R values and softer soils (Site Classes C, D, and E), the transition period is in the range of 3.0 s or higher. For higher R values and firm soils (Classes A and B), the transition period is in the range of 1.0 s, which is applicable for buildings in the range of five to 10 stories.

The basic conclusion from **Fig. G18-2** is this: The minimum base shear, given by Eqs. (12.8-1) and (12.8-5), is likely to control the magnitude of the seismic base shear in areas of moderate to high seismicity, even for low-rise buildings. This is true particularly when the building is situated on Site Class A or B soils.

The absolute lower limit on C_s is 0.01 g. This limit is used only in association with Eq. (12.8-5) [because Eq. (12.8-6) always results in a value greater than 0.01 g]. This limit controls only when $S_{DS} < 0.227/I_e$. For Risk Categories I, II, or III, the importance factor is 1.0 or 1.25, giving limiting values of S_{DS} of 0.227 or 0.181 g, respectively. Thus, according to Table 11.6-1, the lower limit of 0.01 g is applicable only for Seismic Design Category A and B buildings. When the importance factor is 1.5, the limiting value of S_{DS} is 0.151 g, and for Risk Category IV, the Seismic Design Category is A. From this observation, the following statement may be made: The absolute minimum base shear, given by 0.01 W, may control, but only for Seismic Design Category B buildings and lower.

One more general statement may be made for **Fig. G18-2** (and similar figures for different S_1/S_S ratios, not shown herein): Eq. (12.8-4) is never applicable for buildings on Site Class A and B soils and is rarely needed for buildings on Site Class C and D soils because the controlling period is never greater than 4.0 seconds (the minimum value of T_L).

Use of Equations (12.8-4), (12.8-5), and (12.8-6) in Computing Displacements

Eqs. (12.8-5) and (12.8-6) are intended to be used in determining design forces for proportioning members and connections, but not for computing drifts used to check compliance with the drift limits specified in Section 12.12.1. Although it is not clearly stated in Section 12.8.6.2, the intent is that drift may be based on lateral forces developed using the computed period, with or without the upper limit C_uT_a specified in Section 12.8.2. In no circumstances should drift be based on lateral forces computed using Eqs. (12.8-5) or (12.8-6) (i.e., the lower limits on C_s do not apply to drift calculations).

The consequences of using Eq. (12.8-5) in drift computations are shown in the displacement spectra shown in **Fig. G18-3**. The displacement values include the deflection amplifier, $C_d = 5$. The solid line in the figure is the displacement spectrum obtained from Eqs. (12.8-2), (12.8-3), and (12.8-4). The dashed line is based on Eq. (12.8-5). The use of Eq. (12.8-5) produces highly exaggerated displacements at higher periods of vibration. A similar problem occurs when Eq. (12.8-6) controls.

Five-Story Reinforced Concrete Building

In this example, the equivalent lateral forces are determined for a five-story reinforced concrete building. The building houses a health care facility with

Fig. G18-3
Displacement response spectra

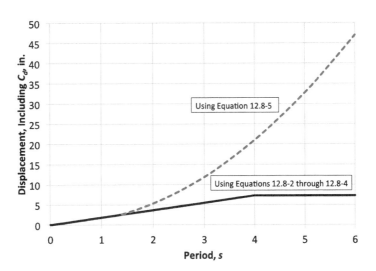

a capacity of more than 50 patients, but no surgical facilities. Pertinent design information is summarized below:

Site Class = B,
$S_S = 0.6g$,
$S_1 = 0.18g$,
$F_a = 1.0$ (from Table 11.4-1),
$F_v = 1.0$ (from Table 11.4-2),
$S_{DS} = (2/3) S_S \times F_a = 0.40g$ [Eqs. (11.4-1) and (11.4-3)],
$S_{D1} = (2/3) S_1 \times F_v = 0.12g$, [Eqs. (11.4-2) and (11.4-4)],
Risk Category = III (from Table 1.5-1),
Importance factor $I_e = 1.25$ (from Table 1.5-2), and
Seismic Design Category = C (from Tables 11.6-1 and 11.6-2).

The ELF method of analysis is permitted for this SDC C building (Table 12.6-1). Based on an SDC of C, an ordinary reinforced concrete shear wall is allowed in the north–south direction and an intermediate reinforced concrete moment frame is allowed in the east–west direction (Table 12.2-1). Design parameters for these systems are as follows:

Ordinary Concrete Shear Wall

$R = 5$,
$\Omega_o = 2.5$,
$C_d = 4.5$, and
Height limit = None.

Intermediate Concrete Moment Frame

$R = 5$,
$\Omega_o = 3$,
$C_d = 4.5$, and
Height limit = None.

A plan of the building is shown in **Fig. G18-4**. The building has a 15-ft-deep basement, a 15-ft-tall first story, and 12-ft-tall upper stories. The slab at grade level is tied into exterior concrete basement walls, and these walls support the moment-resisting frames. The base of the shear walls is interconnected by a grillage of foundation tie beams, and each individual wall is supported by reinforced concrete caissons that extend down to bedrock. Given this detailing, the base of the structure is at grade level, and the basement has no influence on the structural analysis.

The first step in the analysis is to determine the period of vibration. This period, which is different in the two directions because of the use of different structural systems, is calculated using Eq. (12.8-7):

$$T_a = C_t h_n^x \qquad \text{(Eq. 12.8-7)}$$

For both directions, the height, h_n, is taken as the height above grade:

$$h_n = 15 + 4(12) = 63 \text{ ft}$$

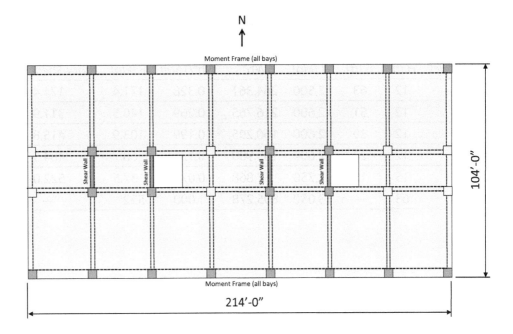

For the shear wall system, Table 12.8-2 provides (in "All other structural systems") $C_t = 0.02$ and $x = 0.75$, giving

$$T_a = 0.02 \times 63^{0.75} = 0.45 \text{ s}$$

For the reinforced concrete moment frame, $C_t = 0.016$ and $x = 0.9$, and

$$T_a = 0.016 \times 63^{0.9} = 0.67 \text{ s}$$

At the preliminary design stage, the computed period from finite element analysis is not available. These computed periods are anticipated to exceed the approximate period, and hence the upper limit period $C_u T_a$ may be used. This assumption would need to be verified before proceeding with a final design.

Using Table 12.8-1 with $S_{D1} = 0.12\,g$, $C_u = 1.9 - 2S_{D1} = 1.9 - 2(0.12) = 1.66$. (See also **Fig. GA-3**, which provides interpolation equations for C_u). Thus the periods to be used for analysis are

$T = 0.45 \times 1.66 = 0.75\,\text{s}$ for the shear wall system, and
$T = 0.67 \times 1.66 = 1.11\,\text{s}$ for the moment frame direction.

The equivalent lateral forces are now developed for the shear wall direction in **Table G18-1**. These forces are based on the following effective seismic weight of the structure:

$$W = 13,050\,k$$

Thus, weight is computed in accordance with Section 12.7.2.

Table G18-1   ELF Forces, Shears, and Overturning Moments for North–South (Shear Wall) Direction

Story/Level	H (ft)	h (ft)	W (kip)	Wh^k	Wh^k/Total	F (kip)	Story Shear (kip)	Story OTM (ft-kip)
5	12	63	2,500	264,361	0.328	171.4	171.4	2,056
4	12	51	2,600	216,765	0.269	140.5	311.9	5,799
3	12	39	2,600	160,295	0.199	103.9	415.8	10,788
2	12	27	2,600	105,988	0.132	68.7	484.5	16,602
1	15	15	2,750	57,868	0.072	37.5	522.0	24,432
Totals	63	—	13,050	805,278	1.000	522	—	—

The equation for finding C_s depends on the values of T, T_S, and T_{MF}.

$$T_S = S_{D1} / S_{DS} = 0.30 \text{ s} \qquad \text{(Section 11.4.5)}$$

$$T_{MF} = 22.7 T_s / R = 22.7 \times 0.3/5 = 1.36 \text{ s} \qquad \text{(Eq. G18-4)}$$

$T_S < T < T_{MF}$, so Eq. (12.8-3) controls. Using this equation gives

$$C_s = \frac{S_{D1}}{T\left(\dfrac{R}{I_e}\right)} = \frac{0.12}{0.75\left(\dfrac{5}{1.25}\right)} = 0.040 > 0.01$$

and from Eq. (12.8-1),

$$V = C_s W = 0.040(13,050) = 522 \text{ kip}$$

The last item that is needed is the exponent k, which is used in association with Eqs. (12.8-1) and (12.8-2) to determine the vertical distribution of forces. With T between 0.5 and 2.5 s, $k = 0.5T + 0.75 = 1.125$. (See **Fig. GA-4**, which provides interpolation functions for k).

Eq. (12.8-3) also controls for the moment frame (east–west) direction. Thus, using $T = 1.11$ s,

$$C_s = \frac{S_{D1}}{T\left(\dfrac{R}{I_e}\right)} = \frac{0.12}{1.11\left(\dfrac{5}{1.25}\right)} = 0.027 > 0.01$$

$$V = C_s W = 0.027(13,050) = 352 \text{ kip}$$

$$k = 0.5T + 0.75 = 1.305$$

The ELF story forces, shears, and overturning moments are shown in **Table G18-2**.

It is interesting to determine how the design forces might change if a special moment frame is used in the east–west direction instead of the

Table G18-2　　　ELF Forces, Shears, and Overturning Moments for East–West (Moment Frame) Direction

Story/Level	H (ft)	h (ft)	W (kip)	Wh^k	Wh^k/Total	F (kip)	Story Shear (kip)	Story OTM (ft-kip)
5	12	63	2,500	557,288	0.350	123.1	123.1	1,478
4	12	51	2,600	439,898	0.276	97.2	220.3	4,121
3	12	39	2,600	309,965	0.195	68.5	288.8	7,587
2	12	27	2,600	191,842	0.120	42.4	331.2	11,561
1	15	15	2,670	94,218	0.059	20.8	352	16,841
Totals	63	—	13,050	565,818	1.000	352	—	—

intermediate frame that was investigated previously. In this case, $R = 8$. The transitional period at which Eq. (12.8-5) controls is

$$T_{MF} = 22.7 T_s / R = 22.7 \times 0.3/8 = 0.85 \text{ s}$$

This period is less than the moment frame period of 1.11 s, and hence, Eq. (12.8-5) controls the base shear. This equation gives

$$C_s = 0.044 S_{DS} I_e = 0.044 \times 0.4 \times 1.25 = 0.022 > 0.01$$

Thus, the special moment frame with $R = 8$ is designed for $0.022/0.027 = 0.81$ times the design base shear for the $R = 5$ system, even though the apparent advantage of using $R = 8$ versus $R = 5$ is a design base shear ratio of 5/8 or 0.625. Given the extra costs of detailing a special moment frame compared with an intermediate moment frame, the economic incentive for using the special moment frame has likely disappeared.

Torsion, Amplification of Torsion, and Orthogonal Loading

Accidental torsion must be included in the structural analysis, per Section 12.8.4.2. The building in **Fig. G18-4** probably has a torsional irregularity because of its rectangular shape and the interior location and nonsymmetric layout of the shear walls. If the torsional irregularity does exist, the accidental torsion must be amplified per Section 12.8.4.3, and 3D analysis must be used.

Section 12.5 establishes rules for direction of loading. Section 12.5.3, which pertains to SDC C buildings, requires that orthogonal loading effects be considered if the building has a Type 5 horizontal irregularity. The most practical method for including the orthogonal load requirements is to load the building with 100% of the load in one direction (including accidental torsion) and 30% of the load in the orthogonal direction. The 30% loading is applied without accidental torsion.

According to Section 12.13.4, the overturning forces imparted to the foundation by the shear walls may be reduced 25%. Although such a reduction may also be taken for the moment frame, the implementation of this reduction is not practical.

Two-Stage ELF Procedure per Section 12.2.3.2

Section 12.2.3.2 allows the use of a two-stage equivalent lateral force analysis for structures that have a flexible upper portion over a rigid lower portion, provided the following two criteria are met:

1. The stiffness of the lower portion must be at least 10 times the stiffness of the upper portion.
2. The period of the entire structure shall not be greater than 1.1 times the period of the upper portion considered as a separate structure supported at the transition from the upper to the lower portion.

When the criteria are met, the upper portion is analyzed as a separate structure using the appropriate values of R and ρ for the upper portion, and the lower portion is designed as a separate structure using R and ρ for the lower portion. When analyzing the lower portion, the reactions from the upper portion must be applied as lateral loads at the top of the lower portion, and these reactions must be amplified by the ratio of the R/ρ of the upper portion to the R/ρ of the lower part of the lower portion. This amplification factor must not be less than 1.0. Also, while the upper portion can be analyzed with either equivalent lateral force or modal response spectrum procedures, the bottom portion can only be analyzed with the equivalent lateral force procedure.

The procedure is illustrated for the structure shown in **Fig. G18-5**. The base of the system is an ordinary precast shear wall ($R = 3$), and the upper portion is an ordinary concentrically braced steel frame ($R = 3.25$). The structure is assigned to Seismic Design Category B and thus has $\rho = 1$

Fig. G18-5
Structure analyzed using two-stage ELF method

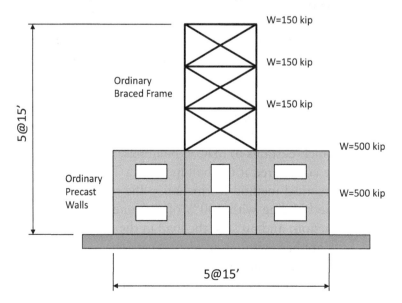

(Section 12.3.4.1). The seismic weights at each level are shown in **Fig. G18-5**. The structure is situated on Site Class B soils. $S_{DS} = 0.25\,g$, and $S_{D1} = 0.072\,g$.

The system was analyzed using SAP 2000 (CSI 2009), wherein the precast walls were modeled with shell elements. The stiffness of the upper portion was determined by fixing it at its base (transition from the upper to the lower portion) and applying a 100-kip lateral load at the top. The displacement at the top was 0.607 in., thus the stiffness is approximately $100/0.607 = 165$ kip/in. The stiffness of the lower portion was found by removing the upper portion and applying a 100-kip lateral force at the top of the lower portion. The displacement at the top of the lower portion was 0.0146 in., and the stiffness is $100/0.0146 = 6{,}872$ kip/in. The ratio of the stiffness of the lower portion to the upper portion is $6{,}872/165 = 41.6$, so the first criterion is met.

The periods of vibration were computed as follows: T for the whole system $= 0.396\,s$, and T for the upper system fixed at the transition from the upper to the lower portion $= 0.367\,s$.

The ratio of the period of the entire structure to the period of the upper portion $= 0.396/0.367 = 1.079$, so the second criterion is met. Given that both criteria are met, the structure may be analyzed using the two-stage ELF method.

The periods of vibration used to determine the period ratios must be determined using a rational analysis and not the approximate formulas provided by Section 12.8.2. Additionally, the stiffness of a structure of several degrees of freedom does not have a unique definition. The approach used previously (wherein a 100-kip load was applied) is only one of several reasonable approaches that might be used.

The equivalent lateral forces for the upper and lower portions of the structure are shown in **Fig. G18-6**. The forces for the upper portion are determined using $W = 450$ kip, $T = 0.367\,s$, and $k = 1.0$. The approximate formula for period (including the $C_u T_a$ limit) is $1.7(0.02 \times 45^{0.75}) = 0.591\,s$. This result is greater than the computed period ($T = 0.367\,s$), so the

Fig. G18-6

Forces on upper and lower systems

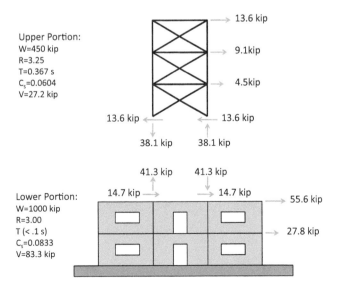

computed period was used. The equivalent lateral forces for the lower portion were determined using $W = 1,000\,k$; Eq. (12.8-2), which always controls for very small T; and $k = 1.0$. When determining the reactions delivered from the upper portion to the lower portion, a magnification factor of $3.25/3 = 1.083$ was used.

When checking drift, the C_d values appropriate for the upper or lower portion should be used. The drift check is not shown herein.

Example 19

Drift and P-Delta Effects

In this example, the drift for a nine-story office building is calculated in accordance with Section 12.8.6 and then checked against the acceptance criteria provided in Section 12.12 and specifically in Table 12.12-1. P-delta effects are then reviewed in accordance with Section 12.8.7. The building is analyzed using the equivalent lateral force method, but only those aspects of the analysis that are pertinent to drift and P-delta are presented in detail.

The seismic force–resisting structural system for the example building is a structural steel moment-resisting space frame, placed at the perimeter. Each perimeter frame has five bays, each 30 ft wide. There is one 12-ft-deep basement level, an 18-ft-high first story, and eight additional upper stories, each with a height of 13 ft. The building has no horizontal or vertical structural irregularities. The total height of the structure above grade is 122 ft. The building is located in Seattle, Washington, on Site Class D soils. An elevation of one of the perimeter frames is shown in **Fig. G19-1**. The beam sizes are the same for each bay across a level, and the column size is the same for all columns in a given story. Doubler plates are used in interior columns only.

Table G19-1 provides the live load weights, the dead load weights, and the total weights for each level of the entire building. Live loads, needed for the P-delta analysis, are based on a reduced live load of 20 psf acting over the full floor. This estimate is sufficient for the purposes of this example or for a preliminary design, but developing more accurate values for a final design is advisable. The seismic weights at each level are equal to the given dead load weights.

In this example, the lateral loads used to compute drift are determined two ways. In the first case, loads are based on the upper limit period of vibration, $T = C_u T_a$, computed in accordance with Section 12.8.2. In the second case, lateral loads are computed using the period of vibration determined from a rigorous (finite element) analysis of the system. This period is

Fig. G19-1

Elevation of building used for P-delta analysis

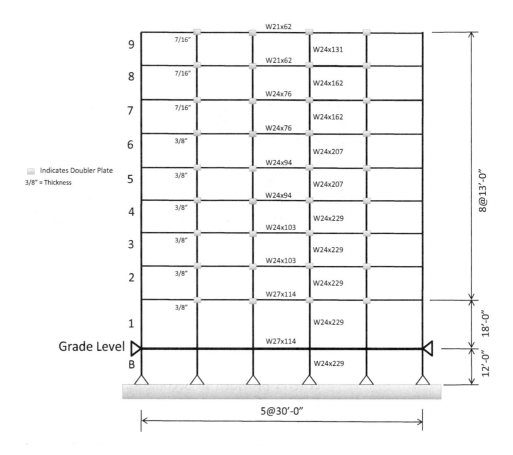

Table G19-1   Building Dimensions and Weights

Story x	Height (in.)	W_Live (kip)	W_Dead (kip)	W_Total (kip)
9	156	450	2,250	2,700
8	156	450	2,325	2,775
7	156	450	2,325	2,775
6	156	450	2,325	2,775
5	156	450	2,325	2,775
4	156	450	2,325	2,775
3	156	450	2,325	2,775
2	156	450	2,325	2,775
1	216	450	2,475	2,925

referred to as $T_{computed}$ in this example. The second case is used to illustrate the potential benefit of Section 12.8.6.2, which allows $T_{computed}$ to be used to determine the lateral loads that are applied to the structure for the purpose of computing drift.

The design spectral accelerations for the Site Class D location are as follows:

$$S_S = 1.25\,g \qquad F_a = 1.0 \qquad S_{DS} = 0.83\,g$$
$$S_1 = 0.5\,g \qquad F_v = 1.5 \qquad S_{D1} = 0.50\,g$$

The Risk Category for the building is III, the importance factor I_e is 1.25, and the Seismic Design Category is D.

Using Table 12.2-1, the response modification coefficient, R, is 8 for the special steel moment frame, and the deflection amplification factor, C_d, is 5.5. There is no height limit for special moment frames.

The effective seismic weight of the entire building, W, is 21,000 kips. The approximate fundamental period of the building is computed using Eq. (12.8-7), with $h_n = 122$ ft and coefficients $C_t = 0.028$ and $x = 0.8$ from Table 12.8-2:

$$T_a = C_t h_n^x = 0.028 \times 122^{0.8} = 1.31 \text{ s}$$

The height used in this computation (122 ft) is based on the assumption that the structure is laterally restrained at the grade level (as shown by a pin support in **Fig. G19-1**).

Using Table 12.8-1, the coefficient for upper limit on period, C_u, is 1.4, thus the period T used for determining base shear and lateral loads is

$$T = C_u T_a = 1.4 \times 1.31 = 1.83 \text{ s}$$

However, the upper limit on period may be used only if a computed period based on a properly substantiated structural analysis is available. The analytical periods were computed using centerline analysis, which approximately accounts for deformations in the panel zones of the beam–column joints, but does not explicitly include the stiffening effect of the doubler plates. The inclusion of panel zone flexibility is required in Section 12.7.3. Section 12.7.3 also requires that the analytical model include P-delta effects, but this is somewhat at odds with the requirements of Section 12.9.6, and specifically Eq. (12.8-16), which uses drift quantities Δ that come from an analysis that *does not* include P-delta effects. The application of Eq. (12.8-16) to results from an analysis that includes P-delta effects would be including such effects twice. This problem is recognized in the last paragraph of Section 12.8.7 and is discussed later in this example.

The computer analysis resulted in $T_{\text{computed}} = 2.95$ s for the model that did not explicitly include P-delta effects. When P-delta effects were explicitly included, the period increased to 3.11 seconds. This 5% increase in period represents approximately a 10% reduction in lateral stiffness when P-delta effects are included.

The computed period is significantly greater than the adjusted period $T = C_u T_a$, and this difference is of some concern. Because of this concern, the computer model was thoroughly checked, and no errors were found. However, differences between the empirical and the computed period, with the computed period greater than the empirical period, are not unusual in moment frame analysis. As shown later in this example, however, periods that are significantly greater than the $C_u T_a$ upper limit may be an indicator that the building is too flexible.

Drift Computations Based on $T = C_u T_a = 1.83$ s

The design seismic base shear based on the upper limit for period of vibration is computed using Eq. (12.8-1):

$$V = C_s W \qquad\qquad (12.8\text{-}1)$$

Equation (12.8-3) controls the value of C_s. Using $T = C_u T_a = 1.83$ s,

$$C_s = \frac{S_{D1}}{T(R/I_e)} = \frac{0.5}{1.83 \times (8/1.25)} = 0.0426\, g$$

Using $W = 21,000$ kip for the entire building,

$$V = C_s W = 0.0426 \times 21,000 = 895 \text{ kip}$$

The equivalent lateral force (ELF) story forces for the full building (not a single frame) are shown in column 2 of **Table G19-2**. These forces were computed according to Eqs. (12.8-11) and (12.8-12), with the exponent $k = 1.665$ for $T = 1.83$ s. Application of these forces to the building resulted in the story displacements δ_{xe} shown in column 3 of the table. See Eq. (12.8-15) and Fig. 12.8-2 of ASCE 7 for a description of symbols used in computing drift.

Interstory drifts (the displacement at the top of a story minus the displacement at the bottom of the same story) are shown in column 4 of **Table G19-2**. The design-level interstory drifts, Δ_x, computed according to Eq. (12.8-15), are shown in column 5. These drifts are based on $C_d = 5.5$ and $I_e = 1.25$.

According to Table 12.12-1 and Section 12.12.1, the interstory drift limits for this Risk Category III building are $0.015/\rho$ times the story height

Table G19-2   Drift Analysis Using $T = C_u T_a$ w/o P-Delta Effects

Story x	F_x (kip)	δ_{xe} (in.)	Δ_{xe} (in.)	$\Delta_x = C_d \Delta_{xe}/I_e$ (in.)	Limit (in.)	Ratio 5/6	Okay?
9	216.5	6.721	0.499	2.195	2.34	0.938	OK
8	185.4	6.222	0.646	2.843	2.34	1.215	NG
7	150.1	5.576	0.752	3.308	2.34	1.414	NG
6	117.8	4.824	0.786	3.458	2.34	1.478	NG
5	88.8	4.038	0.809	3.559	2.34	1.521	NG
4	63.0	3.230	0.818	3.600	2.34	1.538	NG
3	41.0	2.411	0.813	3.575	2.34	1.528	NG
2	22.9	1.599	0.755	3.323	2.34	1.420	NG
1	9.9	0.844	0.844	3.712	3.24	1.146	NG

where ρ is the redundancy factor, which is equal to 1.0 for this highly redundant configuration. These limiting interstory drift values are shown in column 6 of **Table G19-2**. The ratio of the design-level drift to the drift limit is provided in column 7. The limits are exceeded (ratios greater than 1.0) at all stories except for story 9. In story 4, the computed drift is 1.538 times the specified limit.

Drift Computations Based on $T = T_{computed} = 2.95\,\text{s}$

The calculations are now repeated for the same structure analyzed with lateral forces consistent with the computed period of 2.95 s, which is the period computed when P-delta effects are *not* explicitly included in the analysis.

The seismic coefficient C_s for $T = 2.95\,\text{s}$ is

$$C_s = \frac{S_{D1}}{T(R/I_e)} = \frac{0.5}{2.95 \times (8/1.25)} = 0.0265\,g$$

Using $W = 21{,}000\,\text{kip}$ for the entire building,

$$V = C_s W = 0.0265 \times 21{,}000 = 556\,\text{kip}$$

The results of the drift analysis are presented in **Table G19-3**. The exponent k for computing the distribution (see Section 12.8-3) of lateral forces is 2.0 in this case. The drifts have reduced substantially and do not exceed the limiting values at any level. In this case, using the computed period when calculating drifts appears to provide a significant advantage. However, a drift analysis is not complete without performing a P-delta

Table G19-3

Drift Analysis Using $T = T_{computed}$ w/o P-Delta Effects

Story x	F_x (kip)	δ_{xe} (in.)	Δ_{xe} (in.)	$\Delta_x = C_d \Delta_{xe}/I_e$ (in.)	Limit (in.)	Ratio 5/6	Okay?
9	148.9	4.337	0.338	1.486	2.34	0.635	OK
8	122.9	3.999	0.434	1.908	2.34	0.815	OK
7	95.3	3.556	0.498	2.191	2.34	0.936	OK
6	71.3	3.067	0.514	2.260	2.34	0.966	OK
5	50.7	2.554	0.522	2.295	2.34	0.981	OK
4	33.6	2.032	0.521	2.294	2.34	0.980	OK
3	20.0	1.511	0.513	2.255	2.34	0.964	OK
2	9.9	0.998	0.473	2.080	2.34	0.889	OK
1	3.5	0.526	0.526	2.312	3.24	0.714	OK

check. This check is performed in the following section for the case where $T_{computed} = 2.95$ s is used to determine the story forces used in drift analysis.

P-Delta Effects

The P-delta check is carried out in accordance with Section 12.8.7. In the P-delta check, the stability ratio is computed for each story, in accordance with Eq. (12.8-16):

$$\theta = \frac{P_x \Delta I_e}{V_x h_{sx} C_d}$$ (Eq. 12.8-16)

where

P_x = the total vertical design gravity load at level x,

Δ = the interstory drift at level x and is based on center of mass story displacements computed using Eq. (12.8-15),

I_e = the seismic importance factor,

V_x = the total design shear at level x and is based on C_s computed using Eq. (12.8-3) and thus includes the importance factor as a multiplier,

h_{sx} = the story height, and

C_d = the deflection amplifier from Table 12.2-1.

The results of the P-delta analysis are shown in **Table G19-4**. Column 3 of the table provides the accumulated story gravity forces, P, in each story of the building. The gravity forces are unfactored, in accordance with the definition of P_x in Section 12.8.7. The shears in column 4 of **Table G19-4** are the accumulated story shears, while the interstory drifts in column 5 of

Table G19-4 Stability Analysis Using $T = T_{computed}$ and Results from **Table G19-3**

Story	Height (in.)	P_{total} (k)	V_{story} (k)	Δ_x (in.)	θ	θ_{max}	Ratio	OK?	Required Overstrength $= 1/\beta$
9	156	2,700	148.9	1.486	0.039	0.091	0.429	OK	1.0
8	156	5,475	271.8	1.908	0.056	0.091	0.615	OK	1.0
7	156	8,250	367.1	2.191	0.072	0.091	0.791	OK	1.0
6	156	11,025	438.4	2.260	0.082	0.091	0.901	OK	1.0
5	156	13,800	489.1	2.295	0.094	0.091	1.033	NG	1.033
4	156	16,575	522.7	2.294	0.106	0.091	1.165	NG	1.165
3	156	19,450	542.7	2.255	0.118	0.091	1.296	NG	1.296
2	156	22,125	552.6	2.080	0.121	0.091	1.330	NG	1.330
1	216	25,050	556.1	2.312	0.110	0.091	1.208	NG	1.208

Table G19-4 are the same as those in column 5 of **Table G19-3**. The calculated stability ratios are in column 6 of **Table G19-4**, where the maximum value of 0.121 occurs at level 2.

The limiting value of θ is given by Eq. (12.8-17):

$$\theta_{max} = \frac{0.5}{\beta C_d}$$

where β is the ratio of shear demand to shear capacity of the story, which in essence is the inverse of the story overstrength. If β is taken as 1.0, θ_{max} is 0.5/5.5 = 0.091 and is the same for all stories. **Table G19-4** shows that the ratio of 0.091 is exceeded at stories 1 through 5. Without further analysis the building would be deemed to be noncompliant with the stability requirements and would have to be redesigned. However, the redesign can be avoided if the story overstrengths can be shown to be greater than the ratio values shown in column 8 of **Table G19-4**. For example, the required overstrength for story 2 is 1.330.

The story overstrength is likely significantly greater than 1.330 for story 2 because of the strong column–weak beam rules that are built into the various design specifications, such as the *Seismic Provisions for Structural Steel Buildings* (AISC 2010b). Many other factors would also contribute to overstrength, such as actual versus nominal yield strength, strain hardening, and plastic hinging sequence. As shown later in the example, the computed overstrengths for the structure are more than sufficient to satisfy the stability requirements (θ < θ_{max}) for this structure. However, the influence of P-delta effects on story drift must still be investigated because the computed θ values are greater than 0.10 at most stories.

As stipulated in Section 12.8.7, the drifts adjusted for P-delta effects are determined by multiplying the drifts computed without P-delta by the quantity 1/(1 − θ). **Table G19-5** shows the P-delta adjusted story drifts. The drifts shown in column 2 of the table are the same as those shown in column

Table G19-5 P-Delta–Adjusted Drifts from **Table G19-3**

Story x	$\Delta_x = C_d \Delta_{xe}/I_e$ (in.)	θ	$\Delta_x/(1- \theta)$	Limit (in.)	Ratio 4/5	Okay?
9	1.486	0.039	1.546	2.34	0.661	OK
8	1.908	0.056	2.021	2.34	0.864	OK
7	2.191	0.072	2.361	2.34	1.009	NG
6	2.260	0.082	2.461	2.34	1.052	NG
5	2.295	0.094	2.533	2.34	1.082	NG
4	2.294	0.106	2.566	2.34	1.096	NG
3	2.255	0.118	2.523	2.34	1.078	NG
2	2.080	0.121	2.366	2.34	1.011	NG
1	2.312	0.110	2.569	3.24	0.793	OK

Table G19-6   Drift Analysis of Building in **Fig. G19-1** Using $T = T_{computed}$ with P-Delta Effects

Story x	F_x (kip)	δ_{xe} (in.)	Δ_{xe} (in.)	$\Delta_x = C_d\Delta_{xe}/I_e$ (in.)	Limit (in.)	Ratio 7/8	Okay?
9	141.2	4.535	0.339	1.491	2.34	0.637	OK
8	116.6	4.196	0.439	1.932	2.34	0.826	OK
7	90.4	3.758	0.510	2.244	2.34	0.958	OK
6	67.6	3.247	0.532	2.341	2.34	1.000	OK
5	48.1	2.715	0.547	2.407	2.34	1.028	NG
4	31.9	2.168	0.552	2.429	2.34	1.065	NG
3	19.0	1.616	0.548	2.411	2.34	1.030	NG
2	9.4	1.068	0.507	2.231	2.34	0.953	OK
1	3.4	0.561	0.561	2.468	3.24	0.762	OK

5 of **Table G19-3**. Column 4 of **Table G19-5** presents the amplified drifts, which exceed the drift limit at stories 2 through 7.

It is of some interest to perform one additional set of drift calculations for the system analyzed with P-delta effects explicitly included. As mentioned earlier, the period of vibration for this case is 3.11 s. Using this period the base shear is computed as follows:

$$C_s = \frac{S_{D1}}{T(R/I_e)} = \frac{0.5}{3.11 \times (8/1.25)} = 0.0251$$

$$V = C_s W = 0.0251 \times 21,000 = 527 \text{ kip}$$

Table G19-6 shows the resulting drift analysis, where the lateral forces in column 2 are based on $V = 527$ kip and $k = 2.0$. The allowable drift ratios at a few levels are marginally greater than the limits, and thus the drift is not acceptable. The main reason for the difference between the drifts in **Tables G19-5** and **G19-6** is that the values in **Table G19-6** are based on a reduced lateral load of 527 kip (using a computed period of 3.11 s.), versus 556 kip (using a computed period of 2.95 s.).

Backcalculation of Stability Ratios when P-Delta Effects Are Included in Analysis

Many structural analysis programs provide the option to directly include P-delta effects. If an analysis is run with and without P-delta effects, the story stability ratios may be estimated from the results of the two analyses. Although the approach may be used for three-dimensional analysis in theory, the most straightforward use is for two-dimensional analysis. This approach is demonstrated here.

This method is based on the following equation:

$$\Delta_f = \frac{\Delta_0}{1 - \dfrac{P\Delta_0}{VH}} = \frac{\Delta_0}{1 - \theta}$$

(Eq. G19-1)

where

Δ_f = the story drift from the analysis including P-delta effects,
Δ_0 = the drift in the same story for the analysis without P-delta effects,
P = vertical design load in the story [the same as that used in Eq. (12.8-16)],
V = the seismic story shear [the same as that used in Eq. (12.8-16)], and
H = the story height (using the same length units as those used for drift).

The story drifts must be computed from the same lateral loads that produce the story shears, and the story shears must be the same for each analysis. A rearrangement of terms in Eq. (G19-2) produces the simple relationship

$$\theta = 1 - \frac{\Delta_0}{\Delta_f}$$

(Eq. G19-2)

Eq. (G19-2) is demonstrated through the use of data provided in **Table G19-7**. Column 2 of this table provides the story drifts from **Table G19-3**, and column 3 provides the drifts calculated using the same loading and model, except that P-delta effects are included. The stability ratios shown in column 4 of **Table G19-7** were computed using these drifts and Eq. (G19-2). As may be seen, the ratios are very similar to those computed using Eq. (12.8-16).

Table G19-7   Stability Ratios Backcalculated from Analysis Including P-Delta Effects

Story	Δ_0 (in.)	Δ_f (in.)	θ
9	0.338	0.357	0.053
8	0.434	0.463	0.062
7	0.498	0.538	0.074
6	0.514	0.561	0.084
5	0.522	0.576	0.097
4	0.521	0.582	0.105
3	0.513	0.577	0.111
2	0.473	0.534	0.114
1	0.526	0.592	0.111

Computation of Actual Story Overstrengths

As shown earlier in the example, the computed stability ratios θ exceed the maximum allowable value θ_{max} at several levels of the structure. However, the maximum allowable stability ratios were based on β values of 1.0 for each story. Given that β represents the design strength of a story divided by the actual strength, the inverse of β is in fact a measure of overstrength, which is defined as the ratio of actual strength to design strength. To find the value of overstrength $(1/\beta)$ required to satisfy Eq. (12.8-17), the following formula is useful:

$$\frac{1}{\beta} = \frac{C_d \theta_{computed}}{0.5} \qquad \text{(Eq. G19-3)}$$

where $\theta_{computed}$ is the value determined from Eq. (12.8-16). For example, using story 2 from **Table G19-4** where $\theta_{computed} = 0.121$,

$$\frac{1}{\beta} = \frac{5.5 \times 0.121}{0.5} = 1.33$$

This quantity is exactly equal to the ratio shown in column 8 for story 2 in **Table G19-4**, and in fact, these ratios can be used in lieu of Eq. (G19-3). However, a value of $1/\beta$ less than 1.0 would not be used, the minimum reasonable value of $1/\beta$ is 1.0. Given this, the required values of $1/\beta$ required to satisfy Eq. (12.8-17) are shown in column 10 of **Table G19-4**.

Unfortunately, the calculation of actual story strengths is not straightforward and typically requires a series of nonlinear static analyses. A simplified method for estimating story strengths is provided in Section C3 of the commentary to the *Seismic Provisions for Structural Steel Buildings* (AISC, 2005). If the structure has been designed in accordance with the strong column–weak beam design rules, the plastic story strength may be estimated from the following equation:

$$V_{yi} = \frac{2\sum_{j=1}^{n} M_{pGj}}{H} \qquad \text{(Eq. C3-2, AISC, 2005)}$$

where M_{pGj} is the plastic moment capacity of the girder in bay j, n is the number of bays, and H is the story height under consideration.

Using the section sizes shown in **Fig. G19-1** and assuming a yield stress of 50 kip/in.^{2} for steel, the story strengths for one frame are computed as shown in column 5 of **Table G19-8**. Column 6 lists the strength demands, which are based on the story force values in column 2 of **Table G19-2**, but divided by 2.0 to represent a single frame.

Before calculating the true overstrength, the values in column 6 of **Table G19-8** must be divided by the quantity $(1 - \theta)$, as required by Section 12.8.7. The ratio of the computed capacity to the strength demand is shown in

Table G19-8 Story Overstrength Requirements Using Beam Mechanism

Story	θ	θ_{max}	Required Overstrength	V_y (kips)	V_{demand} (kips)	$V_{demand}/(1-\theta)$ (kips)	Ratio 5/7
9	0.039	0.091	1	461	108	113	4.08
8	0.056	0.091	1	461	201	213	2.29
7	0.072	0.091	1	641	276	297	2.16
6	0.082	0.091	1	641	335	365	1.75
5	0.094	0.091	1.033	814	379	418	1.94
4	0.106	0.091	1.165	814	411	459	1.77
3	0.118	0.091	1.296	897	431	488	1.83
2	0.121	0.091	1.330	897	443	504	1.77
1	0.110	0.091	1.208	691	447	502	1.38

column 8 of **Table G19-8**. Clearly, the ratios all exceed the required ratios (1/β values) shown in column 4 of **Table G19-8**. Hence, the structure satisfies the stability requirements of ASCE 7-10 and satisfies the drift requirements when an explicit P-delta analysis is used to compute drifts. Drifts are not satisfied for the structure when the drifts computed without P-delta effects are amplified by the quantity $1/(1 - \theta)$.

The AISC formula, Eq. (C3-2) does not work for braced frames, dual systems, or any other type of structure except a moment frame. Calculating story strengths for general structural systems is not straightforward and may not even be possible without a detailed nonlinear static pushover analysis. For this reason (and several reasons not discussed here), future versions of ASCE 7 are likely to abandon Eq. (12.8-17) in favor of requiring the designer to demonstrate stability through the use of a nonlinear static pushover analysis.

Example 20

Modal Response Spectrum Analysis

In this example, the modal response spectrum (MRS) method of analysis is used to analyze a six-story moment-resisting frame in accordance with Section 12.9 of ASCE 7. The results from the MRS analysis are compared with the results obtained from an equivalent lateral force (ELF) analysis of the same frame. In most cases, an MRS analysis is carried out using commercial finite element analysis software. For this example, however, the analysis is performed using a set of MathCAD routines that have been developed by the author. The use of MathCAD provides details of the analysis that are not easily extracted from the commercial software.

The building in this example is located in Savannah, Georgia. The site for this building is the same as that used in Examples 4 and 5. The building is six stories tall and is used for business offices. According to the descriptions in Table 1.5-1, the Risk Category is II, and from Table 1.5-2, the importance factor I_e is 1.0. Pertinent ground motion parameters are summarized below:

Site Class = D,
S_{DS} = 0.323 g, and
S_{D1} = 0.186 g.

Given these parameters and a Risk Category of II, the building is assigned to Seismic Design Category C. The structural system for the building is an intermediate steel moment frame. According to Table 12.2-1, such systems are allowed in Seismic Design Category C, and they have no height limit. The relevant design parameters for the building are

$R = 4.5$, and
$C_d = 4$.

Fig. G20-1 shows plan and elevation drawings of the building. Moment-resisting frames are placed along lines 1 and 6 in the east–west direction and on lines C and E in the north–south direction. The frames that resist loads in the east–west direction have a series of setbacks, as shown in the south elevation. The frames that resist loads in the north–south direction do not have setbacks. This example considers only the analysis of the frame with setbacks and, more specifically, the frame on gridline 1.

Because of the setbacks, the structure has both a weight irregularity and a vertical geometric irregularity, as described in Table 12.3-2. A preliminary analysis of the complete structure indicates that the structure does not have a torsional irregularity.

According to Table 12.6-1, the equivalent lateral force method may be used to analyze this Seismic Design Category C building. However, because of the vertical irregularities, the modal response spectrum approach is used instead. The details of the MRS method are presented in this example, and the results are compared with those obtained from an ELF analysis of the same system.

The story heights and seismic weights for the frame are provided in Table G20-1. The weights represent the effective seismic mass that is resisted by the frame on gridline 1 only, and they thus represent *one-half* of the mass of the building.

Modeling of Structural System

Section 12.7.3 requires that a three-dimensional model be used whenever the structure has a Type 1a (torsional) or Type 1b (extreme torsional) horizontal irregularity. Such an irregularity does not exist in this case, so analyzing the structure separately in each direction is acceptable.

Note, however, that accidental torsion must be considered and would typically be applied using a static analysis of a three-dimensional model. Given that the purpose of this example is to demonstrate the basic steps involved in modal response spectrum analysis, a two-dimensional model is used in lieu of a more realistic three-dimensional model. Accidental torsion is not included in the analysis presented in this chapter, but is discussed in some detail in Example 14.

For this six-story frame, the analytical model includes 37 nodes. The columns are assumed to be fixed at their base, leaving 30 unrestrained nodes. With three degrees per node (two translations and a rotation) the frame has 90 degrees of freedom (DOF). If it is further assumed that the floor diaphragms are rigid in their own plane, the number of DOF reduces to 66 (30 nodes with two degrees of freedom [one vertical and one rotational], plus six lateral DOF, one for each story).

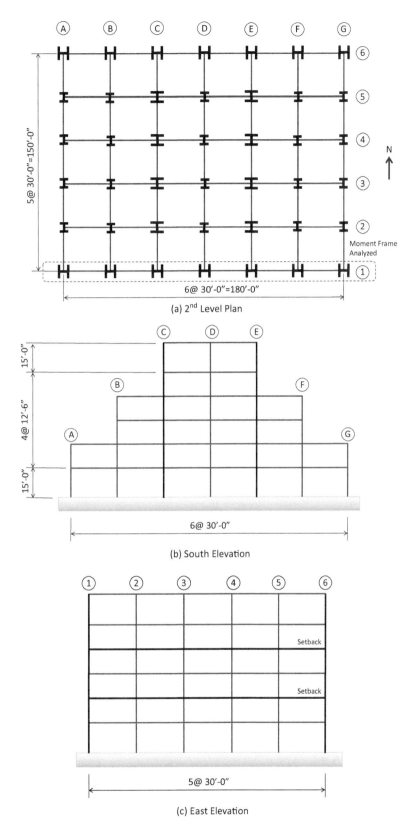

(a) 2nd Level Plan

(b) South Elevation

(c) East Elevation

Table G20-1   Story Properties for Moment Frame

Story	Height (ft)	Seismic Weight[a] (kip)	Seismic Mass[a] (kip-s²/in.)
6	15.0	500	1.294
5	12.5	525	1.359
4	12.5	1,000	2.588
3	12.5	1,025	2.653
2	12.5	1,500	3.882
1	15.0	1,525	3.947
Total	80.0	6,075	15.722

[a]Weights and masses are for half of the full building.

Fig. G20-2
Degrees of freedom
used in MRS analysis

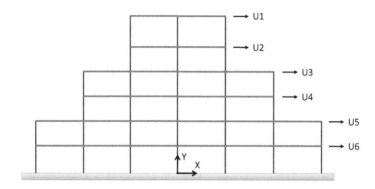

For dynamic analysis, only six degrees of freedom need to be considered, where each degree of freedom represents the lateral displacement in the X-direction at a floor level. As shown in **Fig. G20-2**, these DOF are numbered from the top down. However, the computer model used in the analysis includes all 66 DOF, and the frame's stiffness matrix was assembled on this basis. A six-DOF dynamic stiffness matrix K, shown below, was obtained from the larger matrix by static condensation. Units of all terms are kip/in.

$$K = \begin{bmatrix} 128.72 & -192.46 & 72.98 & -10.48 & 1.44 & -0.22 \\ -192.46 & 521.94 & -410.20 & 91.89 & -12.92 & 1.99 \\ 72.98 & -410.20 & 885.79 & -714.80 & 191.61 & -28.39 \\ -10.48 & 91.89 & -714.80 & 1,352.60 & -896.57 & 198.49 \\ 1.44 & -12.92 & 191.61 & -896.57 & 1,822.86 & -1,374.71 \\ -0.22 & 1.99 & -28.39 & 198.49 & -1,374.71 & 2,260.02 \end{bmatrix}$$

A lumped mass idealization is used, resulting in the diagonal structural mass matrix, M, shown below. This matrix is based on the story weights for a single frame as shown in **Table G20-1**. The units of the terms of the mass matrix are kip-s²/in.

$$M = \begin{bmatrix} 1.29 & 0 & 0 & 0 & 0 & 0 \\ 0 & 1.36 & 0 & 0 & 0 & 0 \\ 0 & 0 & 2.59 & 0 & 0 & 0 \\ 0 & 0 & 0 & 2.65 & 0 & 0 \\ 0 & 0 & 0 & 0 & 3.88 & 0 \\ 0 & 0 & 0 & 0 & 0 & 3.95 \end{bmatrix}$$

Determination of Modal Properties

The next step in the analysis is to determine the modal properties for the system. These properties include
   1) The modal circular frequencies, ω, and periods, T;
   2) The mode shapes, ϕ;
   3) The modal participation factors, Γ; and
   4) The effective modal mass, m.
The mode shapes, ϕ, and modal frequencies, ω, are obtained by solving the eigenvalue problem

$$K\phi = \omega^2 M\phi \qquad \text{(Eq. G20-1)}$$

This equation has six solutions, one mode shape vector ϕ, and one circular vibration frequency ω for each dynamic degree of freedom in the system. MathCAD (PTC, 2012) was used to determine the mode shapes and frequencies, with the results shown as follows. The individual mode shapes are stored column wise in Φ, and the individual circular frequencies (in units of radians per second) are stored along the diagonal of Ω. The mode shapes are normalized such that the maximum ordinate in each mode is exactly 1.0. The mode shapes carry no physical units. The first three modes are plotted in **Fig. G20-3**.

$$\Phi = \begin{bmatrix} \phi_1 & \phi_2 & \phi_3 & \phi_4 & \phi_5 & \phi_6 \end{bmatrix}$$
$$\begin{bmatrix} 1.000 & 1.000 & -0.948 & -0.453 & -0.329 & -0.108 \\ 0.804 & 0.234 & 0.654 & 1.000 & 1.000 & 0.405 \\ 0.615 & -0.300 & 1.000 & -0.052 & -0.777 & -0.591 \\ 0.444 & -0.449 & 0.058 & -0.599 & 0.494 & 1.000 \\ 0.264 & -0.410 & -0.798 & 0.081 & 0.314 & -0.922 \\ 0.131 & -0.235 & -0.640 & 0.384 & -0.640 & 0.781 \end{bmatrix}$$

$$\Omega = \begin{bmatrix} 3.359 & & & & & \\ & 7.143 & & & & \\ & & 11.996 & & & \\ & & & 20.574 & & \\ & & & & 26.236 & \\ & & & & & 32.458 \end{bmatrix}$$

Fig. G20-3
Mode shapes for first
three modes

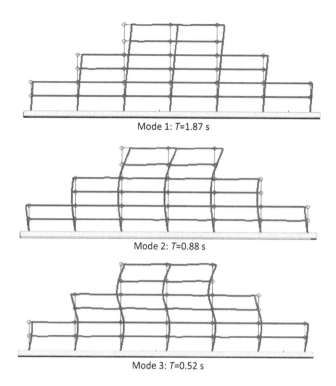

Mode 1: T=1.87 s

Mode 2: T=0.88 s

Mode 3: T=0.52 s

The modal participation factors for each mode i are computed as follows:

$$\Gamma_i = \frac{\phi_i^T M r}{\phi_i^T M \phi_i} \qquad \text{(Eq. G20-2)}$$

where the T superscript on ϕ indicates a matrix transpose, and the term r is a column matrix containing a value of 1.0 in each row. This represents the fact that a horizontal acceleration at the base of the structure directly imposes a horizontal inertial force at each level of the structure.

The effective modal masses are then obtained from the following equation:

$$m_i = \Gamma_i^2 \phi_i^T M \phi_i \qquad \text{(Eq. G20-3)}$$

The modal properties are summarized in **Table G20-2**. (These properties are typically reported by commercial software.) Column 3 contains the modal periods, which are each equal to $2\pi/\omega$. Column 6 contains the accumulated effective modal masses. When all six modes are included, the accumulated mass is 15.723 kip-s²/in., which, as required by theory, is equal to the total mass in the system. Column 7 contains the accumulated mass divided by the total mass, represented as a percent. Section 12.9.1 of ASCE 7 requires that an MRS analysis must include enough modes to capture at least 90% of the actual mass in each orthogonal direction. Only three modes need be considered for the current analysis because the accumulated effective mass for the first three modes is greater than 90% of the total mass. For brevity, the example proceeds with three modes. However, for a system with only six dynamic DOF, including all modes in the analysis would be more reasonable.

Table G20-2   Modal Properties for One-Half of Six-Story Structure

Mode	ω (rad/s)	T (s)	Γ	m (kip-s²/in.)	Accumulated Effective Modal Mass (kip-s²/in.)	Accumulated Mass/Total Mass (% of Total)
1	3.36	1.871	1.669	11.18	11.18	71.1
2	7.14	0.880	−0.958	2.75	13.93	88.6
3	12.0	0.524	−0.382	1.231	15.16	96.4
4	20.6	0.305	0.275	0.242	15.40	98.0
5	26.4	0.238	−0.188	0.203	15.61	99.3
6	32.5	0.194	0.110	0.114	15.72	100.0

Development of Elastic Response Spectrum

The loading for the system is based on the acceleration response spectrum defined in Section 11.4.5. The spectrum is plotted in **Fig. G20-4** for $S_{DS} = 0.323\,g$ and $S_{D1} = 0.186\,g$. The acceleration values associated with all six modal periods of vibration are listed in columns 3 and 4 of **Table G20-3**. (All six modes are shown for completeness, even though only the first three modes are used in the final analysis.) The spectral values do not include the response modification coefficient R. This term is brought into the analysis later. Also, the system damping, assumed to be 5% critical, is included in the development of the response spectrum and need not be considered elsewhere in this analysis.

The spectral displacements for each mode, S_{di}, are determined from the modal spectral accelerations, S_{ai}, through the use of the formula

$$ S_{di} = \frac{S_{ai}}{\omega_i^2} g = \frac{S_{ai} T_i^2}{4\pi^2} g $$

(Eq. G20-4)

Fig. G20-4
Elastic response spectrum

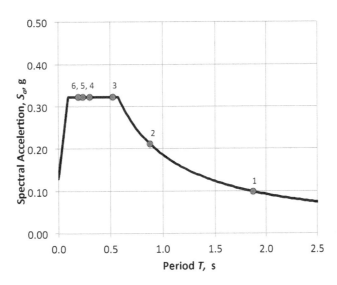

Table G20-3   Spectral Accelerations and Displacements

Mode	Period T (s)	S_a (g)	S_a (in./s^2)	S_d (in.)	ΓS_d (in.)
1	1.871	0.099	38.40	3.403	5.680
2	0.880	0.212	81.64	1.600	−1.531
3	0.524	0.323	124.71	0.867	−0.331
4	0.305	0.323	124.71	0.295	0.081
5	0.238	0.323	124.71	0.179	−0.034
6	0.194	0.323	124.71	0.118	0.013

where g is the acceleration of gravity $= 386.4\,\text{in./s}^2$. The spectral displacements for all six of the system's modes are provided in column 5 of **Table G20-3**.

Column 6 of **Table G20-3** provides the product of the modal participation factor and the spectral displacement for each mode. Because of the scaling of the mode shapes to produce a maximum modal displacement of 1.0 in each mode, the terms in column 6 are a direct indicator of each mode's contribution to the displacements at each floor level of the structure. As expected, the first mode is responsible for most displacement in the system.

Computation of Story Displacements and Story Drift

The next step in the analysis is to determine the displacements in each mode. These displacements are computed as follows:

$$U_i = \Gamma_i S_{di} \phi_i \qquad\qquad \text{(Eq. G20-5)}$$

The displacements are then combined using the square root of the sum of the squares (SRSS) to determine the total displacements at each story. However, story drifts should not be determined from the SRSS of the story displacements. Instead, the drifts should be determined for each mode, and then these story drifts are combined using SRSS. The calculations for story displacement and story drift are provided in **Table G20-4**.

The displacements and story drifts in **Table G20-4** are the elastic values and have not been modified to account for the expected inelastic behavior. To modify for inelastic effects, the values must be multiplied by the quantity (C_d/R). The modified story drifts are provided in **Table G20-5**, together with the limiting values of story drift that are given in Table 12.2-1 of ASCE 7. The story drift limit for this Risk Category II building is 0.02 times the story height. The story drifts appear to be well below the drift limit, particularly in the lower levels.

The displacements in **Tables G20-4** and **G20-5** are based on the computed periods of vibration from the eigenvalue analysis and not on the

Table G20-4   Elastic Displacements and Story Drifts

Story	Elastic Story Displacements (in.)				Elastic Interstory Drifts (in.)			
	Mode 1	Mode 2	Mode 3	SRSS	Mode 1	Mode 2	Mode 3	SRSS
6	5.680	−1.531	0.314	5.891	1.113	−1.173	0.530	1.702
5	4.566	−0.357	−0.217	4.585	1.071	−0.817	0.115	1.352
4	3.496	0.460	−0.331	3.541	0.975	−0.228	−0.312	1.049
3	2.520	0.688	−0.019	2.613	1.023	0.059	−0.283	1.063
2	1.498	0.628	0.264	1.646	0.753	0.268	0.053	0.801
1	0.745	0.360	0.212	0.854	0.745	0.359	0.212	0.854

Table G20-5   Inelastic Story Drifts and Drift Limits

Story	SRSS Drift (in.)	SRSS Drift × (C_d/R) (in.)	Drift Limit (in.)	Ratio (3/4)
6	1.702	1.513	3.6	0.420
5	1.352	1.202	3	0.400
4	1.049	0.932	3	0.311
3	1.063	0.945	3	0.315
2	0.801	0.712	3	0.237
1	0.854	0.759	3.6	0.211

Note: $C_d = 4.0$ and $R = 4.5$.

empirical period T_a or $C_u T_a$. These empirical periods have not yet been computed and are not needed here. Section 12.9.4 allows the drift calculations to be based on the computed period without scaling, which might be required for member forces. This result is consistent with Section 12.8.6.2, which allows the displacements computed by the ELF method to be based on the computed period.

Story Forces and Story Shears

The elastic modal story forces are determined from Eq. (G20-6), in which K is the six-DOF stiffness matrix of the system, and δ_i is the modal story displacement vector for mode i:

$$F_i = KU_i \tag{Eq. G20-6}$$

The elastic story shears are determined for each mode from the elastic story forces, and the total elastic story shears are then computed from the SRSS of the elastic modal story shears. To account for inelastic behavior and for importance, the elastic story shears must be modified by multiplying each value by the ratio (I_e/R) resulting in inelastic story shears. The elastic story forces are shown in **Table G20-6**, and the elastic and inelastic story shears are provided in **Table G20-7**. The inelastic shears are shown in the next to

Table G20-6 Elastic Story Forces

| | Elastic Story Forces (kip) | | |
Story	Mode 1	Mode 2	Mode 3
6	82.9	−101.1	58.4
5	70.0	−24.8	−42.3
4	102.1	60.7	−123.3
3	75.4	93.1	−7.3
2	65.6	124.4	147.7
1	33.2	72.4	120.3

Table G20-7 Story Shears

| | Elastic Story Shears (kip) | | | | Inelastic Story Shear* | Design Story Shears** |
Story	Mode 1	Mode 2	Mode 3	SRSS	(kip)	(kip)
6	82.9	−101.1	58.4	143.2	31.8	42.3
5	152.9	−125.8	16.1	198.7	44.2	58.7
4	255.0	−65.1	−107.2	284.2	63.1	84.0
3	330.4	28.0	−114.5	350.8	78.0	103.7
2	396.0	152.14	33.2	425.6	94.6	125.8
1	429.2	224.8	153.5	508.2	113.0	150.0

*SRSS $\times I_e/R$.

**Inelastic Shear $\times$ 1.33.

the last column of **Table G20-7**. The design story shears, shown in the last column, are based on the scaling requirements of Section 12.9.4 of ASCE 7. The scaling procedure is described in the following text.

The inelastic base shear, which is the same as the first-story inelastic shear (113.0 kip), is computed on the basis of spectral ordinates that are in turn based on the computed periods of vibration for the system. Section 12.9.4 states that the design base shear must not be less than 85% of the base shear computed using the empirical period of vibration, $C_u T_a$.

For a moment-resisting frame, T_a is computed according to Eq. (12.8-7):

$$T_a = C_t h_n^x \tag{Eq. 12.8-7}$$

Using the coefficients for a steel moment frame from Table 12.8-2 and a height of 80 ft,

$$T_a = 0.028(80)^{0.8} = 0.93 \text{ s}$$

Interpolating from Table 12.8-1 with $S_{D1} = 0.186 g$, $C_u = 1.53$, the upper limit on period is

$$C_u T_a = 1.52 \times 0.93 = 1.43 \text{ s}$$

This period is somewhat less than the computed first mode period of vibration, which is 1.87 s. Using Eq. (12.8-1) with $W = 6,075$ kip (for half of the building) and Eq. (12.8-3),

$$C_s = \frac{S_{D1}}{T\left(\dfrac{R}{I_e}\right)} = \frac{0.186}{1.43\left(\dfrac{4.5}{1.0}\right)} = 0.029$$

Checking Eq. (12.8-5), C_s shall not be less than $0.044\, S_{DS} = 0.044(0.323) = 0.0142$, so Eq. (12.8-3) governs, giving

$$V = C_s W = 0.029 \times 6,075 = 176 \text{ kip}$$

According to Section 12.9.4, the design base shear must not be less than 85% of the ELF base shear. For the current example, this is 0.85(176) = 150 kip. This value is greater than the inelastic base shear of 113.0 kip, so the inelastic shears must be scaled by the ratio 150/113.0 = 1.33 to obtain the design shears. The design story shears are shown in the last column of **Table G20-7**.

Computation of Design Member Forces

Elastic member forces are obtained by computing the member forces in each mode and then taking the SRSS of these values, on an element-by-element basis. The elastic member forces are obtained by loading the full 66-DOF system with the elastic modal story forces shown in **Table G20-6**. Design forces are then obtained by multiplying the elastic forces by $I_e/R = 1.0/4.5$, and then (for the current problem) by the scale factor of 1.33.

This procedure is illustrated in **Figs. G20-5(a)** through **G20-5(c)**, which show the modal loading and the resulting elastic member forces.

For the beam (element ID = 15 in **Fig. G20-5**), the combined shears and moments are determined from SRSS as follows:

Elastic moment at left end $= \sqrt{5,020^2 + 1,108^2 + 568^2} = 5,172$ in.-kip;
Elastic moment at right end $= \sqrt{5,005^2 + 1,107^2 + 562^2} = 5,157$ in.-kip;
Elastic shear $= \sqrt{27.9^2 + 6.16^2 + 3.14^2} = 28.7$ kip;
Design moment at left end = 1.33(5,172)(1.0)/(4.5) = 1,529 in.-kip;
Design moment at right end = 1.33(5,157)(1.0)/(4.5) = 1,524 in.-kip; and
Design shear = 1.33(28.7)(1.0)/(4.5) = 8.48 kip.

The values for the column (element ID = 38 in **Fig. G20-5**) are

Elastic moment at bottom $= \sqrt{5,792^2 + 38.4^2 + 2,248^2} = 6,213$ in.-kip;
Elastic moment at top $= \sqrt{5,155^2 + 688^2 + 1,462^2} = 5,402$ in.-kip;
Elastic shear $= \sqrt{73.0^2 + 4.33^2 + 27.4^2} = 78.1$ kip;
Design moment at bottom = 1.33(6,213)(1.0)/(4.5) = 1,836 in.-kip;
Design moment at top = 1.33(5,402)(1.0)/(4.5) = 1,596 in.-kip; and
Design shear = 1.33(78.1)(1.0)/(4.5) = 23.1 kip.

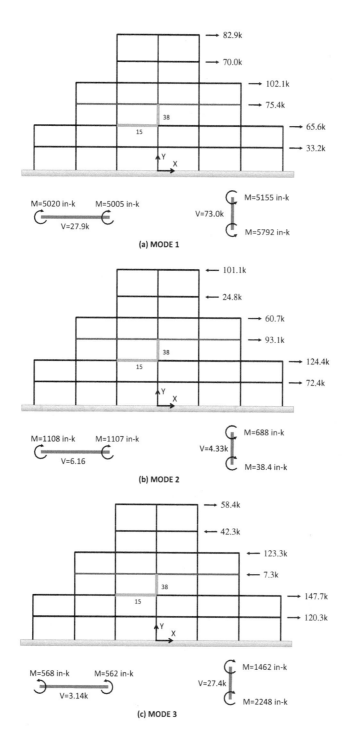

Fig. G20-5
Computing member forces

→ 82.9k
→ 70.0k
→ 102.1k
→ 75.4k
→ 65.6k
→ 33.2k

38
15

Y
X

M=5020 in-k   M=5005 in-k   M=5155 in-k
V=27.9k   V=73.0k
              M=5792 in-k

(a) MODE 1

← 101.1k
← 24.8k
→ 60.7k
→ 93.1k
→ 124.4k
→ 72.4k

38
15

Y
X

M=1108 in-k   M=1107 in-k   M=688 in-k
V=6.16   V=4.33k
              M=38.4 in-k

(b) MODE 2

→ 58.4k
← 42.3k
← 123.3k
← 7.3k
→ 147.7k
→ 120.3k

38
15

Y
X

M=568 in-k   M=562 in-k   M=1462 in-k
V=3.14k   V=27.4k
             M=2248 in-k

(c) MODE 3

Equivalent Lateral Force Analysis

The same frame analyzed with the MRS method was reanalyzed using the ELF method. The results of the analysis are presented in **Tables G20-8** and **G20-9**. Only minimal details are provided for the ELF analysis because the purpose of showing the ELF results is for comparison with the MRS results. A detailed example that considers only the ELF procedure is provided in Example 18.

Table G20-8   ELF Analysis: Story Shears

Story	H (ft)	h_i (ft)	w_i (kip)	$w_ih_i^k$	$w_ih_i^k$/ Total	F (kip)	V (kip)
6	15	80	500	303,422	0.223	39.2	39.2
5	12.5	65	525	235,159	0.173	30.4	69.6
4	12.5	52.5	1,000	327,762	0.240	42.4	112.0
3	12.5	40	1,025	225,722	0.166	29.2	141.2
2	12.5	27.5	1,500	190,972	0.140	24.7	165.9
1	15	15	1,525	80,017	0.059	10.4	176.3
Total	80	—	6,075	1,363,056	1	176.3	—

$k = 1.46$.

Table G20-9   ELF Analysis: Story Drifts

Story	H (ft)	h_i (ft)	w_i (kip)	$w_ih_i^k$	$w_ih_i^k$/Total	F (kip)	$C_d\Delta$ (in.)
6	15	80	500	806,544	0.248	33.3	1.75
5	12.5	65	525	596,794	0.183	24.6	1.61
4	12.5	52.5	1,000	793,101	0.244	32.7	1.38
3	12.5	40	1,025	514,039	0.158	21.2	1.39
2	12.5	27.5	1,500	400,024	0.123	16.5	0.98
1	15	15	1,525	146,410	0.045	6.0	0.94
Total	80	—	6,075	3,256,913	1	134.2	—

$k = 1.68$.

Analysis for story forces and story shears is presented in **Table G20-8**. These forces are based on the upper limit empirical period of vibration $T = C_uT_a = 1.41$ s. As described in the previous section, the design base shear is 176 kip. Distribution of forces along the height is based on Eqs. (12.8-11) and (12.8-12) with the exponent $k = 1.46$ for the period of vibration of 1.41 s. Column 8 of **Table G20-8** contains the design story shears.

Member forces were computed for the structure loaded with the forces in column 7 of **Table G20-8**. The values obtained for the beam and column indicated in **Fig. G20-5** are as follows:

For the Beam

Design moment at left end = –2,126 in.-kip (negative is clockwise),
Design moment at right end = –2,120 in.-kip, and
Design shear = 11.9 kip.

For the Column

> Design moment at bottom = 2,494 in.-kip,
> Design moment at top = 2,195 in.-kip, and
> Design shear = 31.3 kip.

Section 12.8.6.2 states that story drifts may be based on the computed period of vibration. For the system under consideration, the fundamental period from the eigenvalue analysis is 1.871 s, and the computed base shear using this period is 135.6 kip. For computing displacements, lateral forces are obtained using Eqs. (12.8-11) and (12.8-12) with the exponent k based on the computed period. In this case, $k = 1.686$, and the resulting story forces are shown in column 7 of **Table G20-9**. The interstory drifts resulting from the application of these forces to the structure are shown in column 8 of **Table G20-9**. These drifts include the deflection amplification factor $C_d = 4.0$.

The MRS and ELF results are compared in **Tables G20-10, G20-11, and G20-12**. In general, the MRS method produces shears, drifts, and member forces in the neighborhood of 70% of the values obtained using ELF. This difference indicates that for this structure, the use of MRS provides substantial economy, compared with ELF.

Table G20-10   Comparison of MRS and ELF Design Story Shears

Story	MRS Shear (kip)	ELF Shear (kip)	MRS/ELF Shear
6	42.3	39.2	1.08
5	58.7	69.6	0.84
4	84.0	112.0	0.75
3	103.7	141.2	0.73
2	125.8	165.9	0.76
1	150.0	176.3	0.85

Table G20-11   Comparison of MRS and ELF Story Drifts

Story	MRS Drift (in.)	ELF Drift (in.)	MRS/ELF Drift
6	1.51	1.75	0.86
5	1.20	1.61	0.74
4	0.93	1.38	0.67
3	0.94	1.39	0.67
2	0.71	0.98	0.72
1	0.76	0.94	0.81

Comparison of MRS and ELF Member Design Forces

Item	MRS	ELF	MRS/ELF
Beam moment left (in.-kip)	1,529	2,126	0.72
Beam moment right (in.-kip)	1,524	2,120	0.72
Beam shear (kip)	8.48	11.9	0.71
Column moment bottom (in.-kip)	1,836	2,494	0.74
Column moment top (in.-kip)	1,596	2,195	0.73
Column shear (kip)	23.1	31.3	0.74

Three-Dimensional Modal Response Spectrum Analysis

It is beyond the scope of this *Guide* to present a detailed example of a three-dimensional modal response spectrum analysis. However, certain aspects of such an analysis are pertinent, and these aspects are excerpted from an example developed by the author for the "NEHRP Recommended Provisions: Design Examples" document, available on compact disc from FEMA (FEMA P-751, 2012).

The building under consideration is a 12-story steel building with one basement level. The structural system is a perimeter moment-resisting space frame. A three-dimensional wire-frame drawing of the building, as modeled with the SAP 2000 program (CSI 2009) is shown in **Fig. G20-6**. Although

Fig. G20-6
Wire-frame model of 12-story building (lower level is the basement wall)

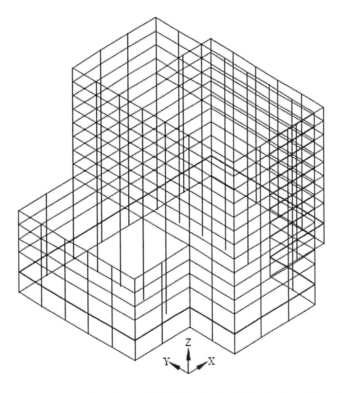

Table G20-13   Modal Properties for 12-Story Steel Moment Frame Building

Mode	Period (s)	Modal Direction Factor		
		X	Y	Torsion
1	2.867	99.2	0.7	0.1
2	2.745	0.8	99.0	0.2
3	1.565	1.7	9.6	88.7
4	1.149	98.2	0.8	1.0
5	1.074	0.4	92.1	7.5
6	0.724	7.9	44.4	47.7
7	0.697	91.7	5.23	3.12
8	0.631	0.3	50.0	49.7
9	0.434	30.0	5.7	64.3
10	0.427	70.3	2.0	27.7

it is not clear from the drawing, the basement level of the building was explicitly modeled, and thick-shell elements were used to represent the basement walls. Floor diaphragms were modeled as rigid in plane and flexible out of plane. For analysis, the structure was assumed to be fixed at the bottom of the basement level. For computing the approximate period T_a (Eq. [12.8-7]), the height of the building would be measured from the top of the basement walls.

The properties of the first 10 mode shapes are shown in **Table G20-13**. For each mode, the periods of vibration and modal direction factors are given. The modal direction factors indicate the predominant direction of the mode. The first mode, with a period of 2.867 s, is a translational response in the X direction. The second mode is translational in the Y direction, and the third mode is torsion. The fourth and fifth modes are predominantly lateral, but the sixth and higher modes have significant lateral–torsional coupling. The first eight mode shapes are plotted in **Fig. G20-7**.

The effective modal masses for the first 10 modes are provided in **Table G20-14**. For each mode, the effective mass for that mode in the direction of interest and the accumulated mass in that direction are given. For the third mode, for example, the effective mass in the X direction is 0.34% of the total mass in the X direction, and the accumulated X direction mass in modes 1, 2, and 3 is 64.9% of the total X-direction mass.

Section 12.9.1 of ASCE 7 requires that enough modes be included to capture at least 90% of the total mass in each direction. Clearly, this requirement is not achieved with 10 modes. In fact, it would take 40 modes to satisfy this requirement; 40 is more modes than one would expect for a 12-story building with rigid diaphragms (10 modes would usually be sufficient). The reason that 10 modes are not sufficient is that almost 18% of the total mass of the structure is represented by the subgrade level of the building, including the basement walls and the grade-level diaphragm. In this sense, the spirit of ASCE 7 could be met with only 10 modes because

Fig. G20-7
First eight mode shapes of a 12-story steel building

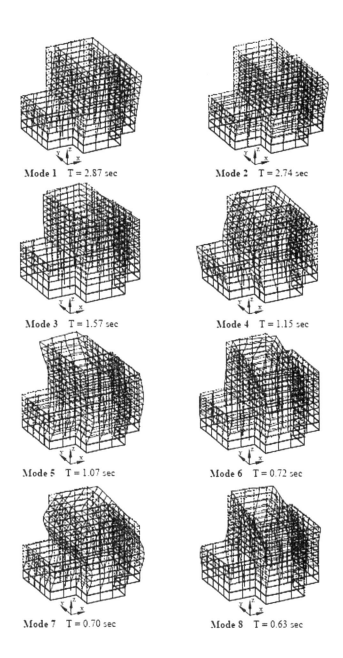

Mode 1   T = 2.87 sec

Mode 2   T = 2.74 sec

Mode 3   T = 1.57 sec

Mode 4   T = 1.15 sec

Mode 5   T = 1.07 sec

Mode 6   T = 0.72 sec

Mode 7   T = 0.70 sec

Mode 8   T = 0.63 sec

these modes capture almost 100% of the dynamically excitable mass in the above-grade portion of the structure. If determining the stresses and forces in the basement walls is desired, at least 40 modes would be required.

Also, the complete quadratic combination method of modal combination should be used for this structure because the midlevel modes have a high degree of lateral–torsional coupling. Modal combination requirements are provided in Section 12.9.3 of ASCE 7.

The SRSS method of combination combines each mode in a probabilistic manner because modes of different periods do not produce maximum responses at the same time. When two modes have very similar periods, however, then their effects should be more correlated with respect to peak response, and the results should be added directly. The CQC method approximately recognizes this and combines the modes more appropriately.

Table G20-14   Effective Modal Masses for 12-Story Steel Moment Frame Building

Mode	X (Mode)	X (Accumulated)	Y (Mode)	Y (Accumulated)	T (Mode)	T (Accumulated)
			Effective Modal Masses (% of Total)			
1	64.04	64.0	0.46	0.5	0.04	0.0
2	0.51	64.6	64.25	64.7	0.02	0.1
3	0.34	64.9	0.93	65.5	51.06	51.1
4	10.78	75.7	0.07	65.7	0.46	51.6
5	0.04	75.7	10.64	76.3	5.30	56.9
6	0.23	75.9	1.08	77.4	2.96	59.8
7	2.94	78.9	0.15	77.6	0.03	59.9
8	0.01	78.9	1.43	79.0	8.93	68.8
9	0.38	79.3	0.00	79.0	3.32	71.1
10	1.37	80.6	0.01	79.0	1.15	72.3

Note: *T* = Torsion.

Example 21

Modal Response History Analysis

In this example, the modal response history (MRH) method of analysis is used to analyze the same six-story moment-resisting frame that was analyzed with the modal response spectrum method and the equivalent lateral force method in Example 20. The MRH analysis is performed in accordance with Section 16.1 of ASCE 7. Analyses are performed with single-period amplitude-scaled records and are repeated with spectrum-matched records. The results from the analyses are compared with the results obtained from the modal response spectrum method and the equivalent lateral force method.

The building analyzed in this example is the same six-story steel moment frame as that used for Example 20, and the site is the same as used for Examples 4, 5, 6, and 20. The present example will first present an overview of the problem and then describe the response history analysis procedure. Much of the work required for the analysis was completed in the previously cited examples. For example, the determination of the basic seismic ground motion parameters S_{DS} and S_{D1} was completed in Example 4, the design response spectrum was constructed in Example 5, and the ground motion selection and scaling process was completed in Example 6. Additionally, all of the required analytical properties (stiffness matrix, mass matrix, mode shapes, frequencies, modal participation factors, etc.) were determined in Example 20. Some of the results from these previous examples are repeated in the current example for ease of readability.

It is convenient to repeat the basic design parameters for the building, as follows:

Location: Savannah Georgia, on Site Class D soils;
Seismic risk parameters: $S_{DS} = 0.323g$, $S_{D1} = 0.186g$;
Risk and importance: Risk Category II, $I_e = 1.0$;
Seismic Design Category: C;
System type: Intermediate steel moment frame; and
System design parameters: $R = 4.5$, $C_d = 4.0$

The plan and elevation drawings of the building are shown in **Fig. G20-1**. Moment-resisting frames are placed along lines 1 and 6 in the east–west direction and on lines C and E in the north–south direction. The frames that resist loads in the east–west direction have a series of setbacks, as shown in the south elevation. The frames that resist loads in the north–south direction do not have setbacks. This example considers only the analysis of the frame with setbacks and, more specifically, the frame on gridline 1.

The story heights and seismic weights for the frame are provided in **Table G20-1**. The weights represent the effective seismic mass that is resisted by the frame on gridline 1 only, and they thus represent one-half of the mass of the building.

Because of the setbacks, the structure has both a weight irregularity and a vertical geometric irregularity, as described in Table 12.3-2. A preliminary analysis of the complete structure indicates that the structure does not have a torsional irregularity.

According to Table 12.6-1, the equivalent lateral force method may be used to analyze this Seismic Design Category C building. However, because of the vertical irregularities, the modal response spectrum method or the linear response history method may be more appropriate. Due to the need to select and scale the ground motions used for analysis, the modal response history method is without doubt more time consuming than is the response spectrum method (particularly for someone who does not have experience working with ground motions). However, the response history approach offers certain advantages, particularly from the perspective that the signs of deflections and member forces are retained. These signs are lost in the response spectrum approach when the modal combinations are made through the square root of the sum of squares (SRSS) procedure.

Modeling of the Structural System

The modeling of the system is described in detail in Example 20. Pertinent to this example is the fact that the modal response history analysis is performed on the reduced, six-degrees-of-freedom (DOF) system, rather than the full 66-DOF system. This is consistent with the reduced system that was used in Example 20 and is illustrated in **Fig. G20-2**. For convenience, the stiffness and mass matrices for the reduced system consisting of *one* of the two frames and *one-half of the mass* of the entire system are shown as follows:

$$K = \begin{bmatrix} 128.72 & -192.46 & 72.98 & -10.48 & 1.44 & -0.22 \\ -192.46 & 521.94 & -410.20 & 91.89 & -12.92 & 1.99 \\ 72.98 & -410.20 & 885.79 & -714.80 & 191.61 & -28.39 \\ -10.48 & 91.89 & -714.80 & 1,352.60 & -896.57 & 198.49 \\ 1.44 & -12.92 & 191.61 & -896.57 & 1,822.86 & -1,374.71 \\ -0.22 & 1.99 & -28.39 & 198.49 & -1,374.71 & 2,260.02 \end{bmatrix}$$

$$M = \begin{bmatrix} 1.29 & 0 & 0 & 0 & 0 & 0 \\ 0 & 1.36 & 0 & 0 & 0 & 0 \\ 0 & 0 & 2.59 & 0 & 0 & 0 \\ 0 & 0 & 0 & 2.65 & 0 & 0 \\ 0 & 0 & 0 & 0 & 3.88 & 0 \\ 0 & 0 & 0 & 0 & 0 & 3.95 \end{bmatrix}$$

The computational units for all terms in the stiffness matrix are kip/in. and for the mass matrix are kip-s²/in.

Determination of Modal Properties

The modal properties for the reduced system were computed in Example 20. The relevant quantities are as follows:
1. The modal frequencies, ω, and periods, T;
2. The mode shapes, ϕ;
3. The modal participation factors, Γ; and
4. The effective modal mass, m.

The computed mode shapes are provided in Example 20, and **Fig. G20-3** illustrates the first three shapes. The important modal properties are presented in **Table G20-2** and are repeated for the reader's convenience in **Table G21-1**.

As mentioned in Example 20, Section 12.9-1 of ASCE-7 requires that enough modes be included in a modal response spectrum analysis to capture no less than 90% of the mass of the system. As seen in column 7 of **Table**

Table G21-1   Modal Properties for One-Half of Six-Story Structure

Mode	ω (rad/s)	T (s)	Γ	m (kip-s²/in.)	Accumulated Effective Modal Mass (kip-s²/in.)	Accumulated Mass/Total Mass (% of Total)
1	3.36	1.871	1.669	11.18	11.18	71.1
2	7.14	0.880	-0.958	2.75	13.93	88.6
3	12.0	0.524	-0.382	1.231	15.16	96.4
4	20.6	0.305	0.275	0.242	15.40	98.0
5	26.4	0.238	-0.188	0.203	15.61	99.3
6	32.5	0.194	0.110	0.114	15.72	100.0

G21-1, this would be accomplished by use of the first three modes, with an accumulated mass participation of 96.4%. ASCE 7 does not provide guidelines for determining the number of modes to be used in modal response history analysis. Using the same number of modes for response spectrum analysis would be reasonable, and thus three modes would be appropriate as a minimum. For the purposes of this example, however, all six modes are used. The influence of this choice on the results of the analysis is briefly described later in the example.

Selection and Scaling of Ground Motions

A detailed description of the ground motion selection process is provided in Chapter 6 and is not repeated here. However, Section 16.1.3.1 requires that all ordinates of the average of the scaled ground motion pseudoacceleration spectra be greater than the ordinates of the target acceleration spectra over a period range from $0.2T$ to $1.5T$, where T is the fundamental period of the structure vibrating in its first mode. The appropriateness of this period range is questionable for linear analysis, particularly with regard to the value of $1.5T$, which is intended to account for yielding and will not occur in a linear analysis. This aspect of the analysis is discussed in summary remarks at the end of this example.

The records chosen for the analysis are summarized in **Table G21-2**, which all represent ground motions recorded on Site Class D soil (the same as the building site) and have magnitudes around 6.5. Epicentral distances greater than 10 km indicate that these motions may be considered far-field.

The scale factors used for the analysis were determined in Example 6 and are summarized in **Table G21-3**. In **Table G21.3** the fundamental period (*FP*) scale factor is required for the individual ground motion spectrum to

Table G21-2   Record Sets Used for Analysis

Earthquake	PEER NGA ID	Year	Magnitude	ASCE 7 Site Class	Fault Type	Epicentral Distance (km)
San Fernando	68	1971	6.6	D	Thrust	39.5
Imperial Valley	169	1979	6.5	D	Strike-slip	33.7
Northridge	953	1994	6.7	D	Thrust	13.3

Table G21-3   Ground Motion Scale Factors Used for Analysis

Ground Motion	FP Scale Factor	S Scale Factor	C Scale Factor = FP × S
1: San Fernando	1.334	1.181	1.575
2: Imperial Valley	0.399	1.181	0.471
3: Northridge	0.305	1.181	0.360

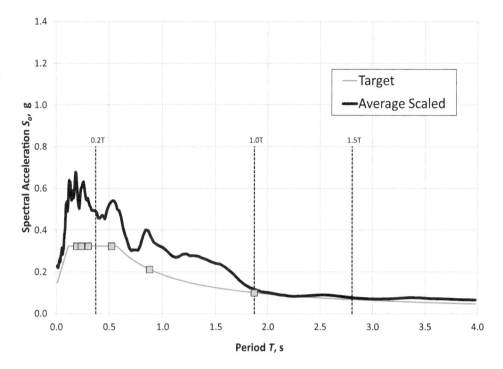

Fig. G21-1

Target spectrum and average amplitude-scaled ground motion spectrum

match the target spectrum at the fundamental period T, the suite (S) scale factor is the additional factor required for the average of the FP scaled spectra to exceed the target spectrum between $0.2T$ and $1.5T$, and the combined (C) scale factor is the product of FP and S for each motion.

The match period for the FP scaling was equal to the fundamental period of the system, $T = 1.87\,\text{s}$, and the scaling range was $0.374\,\text{s}$ to $2.80\,\text{s}$. **Fig. G21-1** shows a plot of the average of the amplitude-scaled spectra together with the target spectrum, and **Fig. G21-2** shows plots of each individual amplitude-scaled spectrum together with the target spectrum. The rectangular symbols on the target spectrum of **Fig. G21-1** represent the modal periods of the building being analyzed. Note that the three lowest periods are less than 0.2 times the fundamental period, and hence, these do not influence the scaling.

Fig. G21-2 shows that the individual scaled spectra are close to the target spectrum at the structure's fundamental period $T = 1.87\,\text{s}$, but for the San Fernando and Northridge records there is significant deviation between the scaled spectra and target spectra at periods less than the fundamental period. This will lead to a greater higher mode response, and thus a greater total response for these records compared with the Imperial Valley earthquake.

For the purpose of this example, the response history analysis is repeated for the ground motion records scaled using a spectrum-matching procedure. In spectrum matching, the original record is decomposed in the frequency domain using Fourier transforms or wavelets, and the decomposed parts of the record are scaled and reassembled such that the spectrum for the modified motion closely matches the design spectrum. For this example the matching was performed using the RSP-Match computer program (Hancock et al., 2006), which utilizes wavelet decomposition. **Fig. G21-3** shows that the response spectra for the spectrum-matched records are very similar in

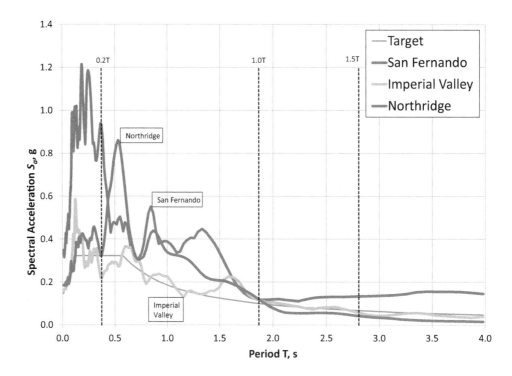

Fig. G21-2
Target spectrum and individual scaled ground motion spectra

shape to the design spectrum. These matched records were *not* amplitude scaled to comply with the requirements of Section 16.1.3.1 because the spectrum-matching approach is considered as an alternate to the approach described in Section 16.1.3.1. The commentary Section C16.1.3.2 specifically allows spectrum matching for three-dimensional analysis, but section C16.1.3.1 does not indicate that spectrum matching is allowed for two-dimensional analysis. The reader is encouraged to see NIST (2011) for more information on the ground motion selection and scaling process.

Overview of Modal Response History Methodology

Once the modal properties are known, the analysis procedure is straightforward. For each mode *i* included in the analysis the following differential equation must be solved:

$$\ddot{y}_i(t) + 2\xi_i\omega_i\dot{y}_i(t) + \omega_i^2 y_i(t) = -\Gamma_i\ddot{u}_g(t) \qquad \text{(Eq. G21-1)}$$

where

ξ_i = the damping ratio in the given mode;

$\ddot{u}_g(t)$ = the amplitude-scaled or spectrum-matched history of ground accelerations (in acceleration units); and

$y_i(t)$ and its derivatives = the histories of modal displacement, velocity, and acceleration. The terms ω_i, and Γ_i represent the modal circular frequency and the modal participation factor, respectively.

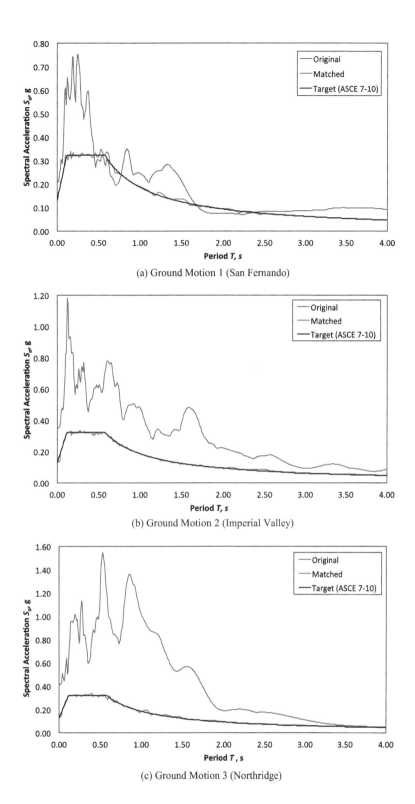

Fig. G21-3

Spectrum-matched ground motion spectra

(a) Ground Motion 1 (San Fernando)

(b) Ground Motion 2 (Imperial Valley)

(c) Ground Motion 3 (Northridge)

ASCE 7 does not specify the level of damping to use in the response history analysis. In this example a modal damping ratio ξ of 0.05 is used in each mode, which is consistent with the damping ratio used in the development of the target and ground motion spectra. Other damping ratios could be used, but this would require an adjustment of the target spectrum for damping other than 5% critical, recalculation of the individual ground motion spectra for the same level of damping, and recalculation of the ground motion scale factors. While a damping ratio of 0.05 is reasonable for linear response history analysis, values in the range of 0.02 to 0.03 are more appropriate for nonlinear analysis where inelastic behavior is explicitly included (NIST, 2010a).

The modal displacement histories $y_i(t)$ in Eq. (G21-1) may be found using several techniques, the most convenient of which is the "interpolation of excitation" method (Chopra, 2011). This method, also called the piecewise exact method, produces a mathematically exact response when the loading history consists of a series of straight-line segments. Ground acceleration histories are always represented by this type of loading because the digitized record of acceleration contains only discrete values, recorded at some constant time increment (usually in the range of 0.005 to 0.02 seconds). All three records used in the analysis were recorded at a time interval of 0.01 seconds. Response history calculations were performed using the same time interval.

Once the scalar modal displacement histories $y_i(t)$ are known, the mode's contribution to the lateral displacement in the reduced (6-DOF) coordinate system are obtained as follows:

$$U_i(t) = \phi_i y_i(t) \qquad \text{(Eq. G21-2)}$$

Note that for the system under consideration, U_i is an array (matrix) consisting of six rows (one for each degree of freedom shown in **Fig. G20-2**) and n columns, where n is the number of individual acceleration values in the ground acceleration history. For the DOF numbering used in this example the first row of U_i represents the history of roof displacement of the structure and the 100th column (for example) represents a "snapshot" of the displacement profile of the structure one second (0.01 seconds times 100 steps) into the response.

The reduced system displacement history is simply taken as the sum of the modal combinations:

$$U(t) = \sum_{i=1}^{\text{modes}} U_i(t) \qquad \text{(Eq. G21-3)}$$

Chapter 16 of ASCE 7 does not specify how many modes should be used in linear response history analysis. With respect to modal response spectrum analysis, Section 12.9.1 states that enough modes should be used such that at least 90% of the effective mass is captured in the direction analyzed. As mentioned in Chapter 20 of this *Guide*, three modes are sufficient, and thus three modes were used in the response history analysis. Preliminary calculations indicated that the use of six modes instead of three modes generally increased the response values by less than 3%.

The displacements $U(t)$ may then be used to compute the interstory drifts at any time during the response. Story inertial forces, $F(t)$ may be obtained by Eq. (G21-4):

$$F(t) = KU(t) \qquad \text{(Eq. G21-4)}$$

where K is the reduced six-by-six stiffness matrix for the system. As with U, F is an array of six rows (one for each degree of freedom) and n columns, where n is the number of individual values in the ground acceleration history. These forces may then be used to determine story shears, base shears, story overturning moments, and the base overturning moment.

These inertial forces may then be applied at the appropriate lateral degrees of freedom of the full system to produce displacement histories for all 66 degrees of freedom in the model. Using these expanded displacements, the individual element force histories may be found for each beam and column of the structure.

Up to this point all displacements and member forces are based on an elastic analysis. For design purposes all forces (member forces, story shears, story overturning moments, and reactions) must be multiplied by I_e/R, and all displacements must be multiplied by C_d/R.

Section 16.1.4 of ASCE 7 requires that all force-related quantities from the analysis be scaled if the computed peak base shear (including the I_e/R factors) is less than 0.85 times the *minimum* equivalent lateral force base shear, where $V = C_s W$ and for the current example (with $S_1 < 0.6\,g$), $C_s = 0.044 S_{DS} I_e \geq 0.01$. Using $S_{DS} = 0.323$ and $I_e = 1.0$, $C_s = (0.044)(0.323)(1.0) = 0.0142$. With W for one frame = 6,075 kip, the minimum base shear for one frame is 86.3 kip. As shown later in the example, design base shear from response history analysis for a single frame is well in excess of 86.3 kips, so member forces need not be scaled. With regards to scaling it is noted that when response spectrum analysis is performed, the computed member forces must be scaled upward if the computed design base shear is less than 0.85 times the base equivalent lateral force base shear, regardless of the C_s used to compute the base shear. The rationale for the difference in scaling requirements between the response spectrum method and the response history method is not clear, and from the author's perspective, forces from the response history analysis should be scaled in a manner consistent with those using response spectrum analysis. For this example, however, the procedures of ASCE 7 were followed.[1]

Section 16.1.4 of ASCE 7 provides the criteria for determining the final design values. If fewer than seven ground motions are used, as in this example, the *maximum* of any quantity computed among these motions must be used. However, signs can be retained when computing the maxima, producing, for example, peak positive moments in beams, peak negative moments in beams, peak tension in braces, and peak compression in braces. These peak values are generally not concurrent among individual ground motions. It is also important to note that in some cases pairs of values must be used. For example, for a column, axial-force and bending-moment pairs are needed. Thus, for such elements, the maximum positive bending moment and the corresponding axial force (regardless of sign and magnitude) must be recorded, as well as the maximum positive axial force and the corresponding bending moment (regardless of sign and magnitude).

Table G21-4 Individual and Peak Interstory Drift for Analysis Using Amplitude Scaled Records

Story	Ground Motion 1		Ground Motion 2		Ground Motion 3		Peak	Design = Peak × C_d/R
	Min	Max	Min	Max	Min	Max		
6	−3.884	2.864	−2.478	1.791	−3.397	3.026	3.884	**3.452**
5	−2.866	2.286	−1.944	1.424	−2.647	2.526	2.886	**2.656**
4	−1.330	1.269	−1.227	1.193	−1.395	1.486	1.486	**1.321**
3	−1.396	0.997	−1.122	1.182	−1.146	0.887	1.396	**1.241**
2	−1.104	0.886	−0.738	0.866	−0.821	0.986	1.104	**0.981**
1	−1.187	1.056	−0.800	0.946	−1.032	1.153	1.187	**1.055**

All displacement in in.

Results Using Amplitude-Scaled Ground Motions

Story Displacements and Story Drift

The minimum and maximum *unscaled* interstory drifts computed for each motion are provided in **Table G21-4**. The peak unscaled drift for each story is also provided, as is the maximum design value (shown in bold text) scaled by C_d/R. The allowable drifts for this Risk Category II building are 2% of the story height. The story drift limit is then $0.02 \times 180 = 3.60$ in. for the 1st and 6th story and $0.02 \times 150 = 3.0$ in. for stories 2, 3, 4, and 5. The computed drifts are lower than the allowable drifts for all stories. Note, however, that the upper level drifts are much higher than the lower level drifts, and that drifts for ground motions 1 and 3 are greater than that for ground motion 2. This is due to higher mode effects accentuated by the large differences between the scaled record acceleration and the target spectrum at the lower periods (see **Fig. G21-2**).

Story Shears, Overturning Moments, and Member Forces

The story shears were computed from the inertial forces, with the results summarized in **Table G21-5**. First, the minimum and maximum unscaled shears are shown at each story for each individual ground motion, and then the peak shears are taken as the maximum magnitude among the three ground motions for each story. Design values are determined by multiplying the peak shears by I_e/R. The values shown for the first story are the same as the base shear. Note that the peak design base shear is significantly greater than the minimum base shear of 86.3 kips and is less than the base shear computed using the ELF procedure, which is 176 kip (see Example 20 of this *Guide*).

Note the large variation in base shear among the three records. In particular, the results for ground motions 1 and 3 are larger than those for

Table G21-5   Individual and Peak Story Shears for Analysis Using Amplitude Scaled Records

		Ground Motion 1		Ground Motion 2		Ground Motion 3			Design =
Story	Min	Max	Min	Max	Min	Max	Peak	Peak ×I_e/R
6	−328.8	234.5	−204.8	148.1	−280.6	241.5	328.8	**73.1**
5	−428.0	344.7	−287.4	210.7	−397.1	381.0	428.0	**95.1**
4	−360.7	356.1	−337.0	316.7	−383.1	421.8	421.8	**93.7**
3	−487.0	338.6	−371.3	380.8	−418.0	321.1	487.0	**108.2**
2	−593.9	486.6	−388.4	462.0	−449.0	539.9	593.9	**131.9**
1	−726.4	644.0	−484.0	566.3	−640.7	698.5	726.4	**161.4**

All shears in kip.

ground motion 2. This is due to the poor matching of the scaled ground motion spectra relative to the design spectra at the lower periods (see **Fig. G21-2**).

Story overturning histories were computed at each level, with results shown in **Table G21-6**. These moments are at the bottom of the indicated story. It is interesting to note that the moments at the upper stories for ground motion 2 are significantly less than for ground motions 1 and 3, but that the moments at the lower stories for ground motion 2 are close to the values for ground motion 1 and somewhat higher than the moments at the lower stories for ground motion 3.

The minimum, maximum, and peak moments and shear in beam element 15 and column element 38 are provided in **Table G21-7**, together with the design value, which is I_e/R times the peak value. Note that one of the advantages of response history analysis is that separate peak values for positive and negative moment can be obtained. For the purpose of this example, the largest positive or negative moment is recorded as the peak, regardless of sign.

Table G21-6   Individual and Peak Story Overturning Moments for Analysis Using Amplitude Scaled Records

	Ground Motion 1		Ground Motion 2		Ground Motion 3			Design =
Story	Min	Max	Min	Max	Min	Max	Peak	Peak ×I_e/R
6	−4,931	3,518	−3,073	2,223	−4,209	3,623	4,941	**1,098**
5	−10,272	7,788	−6,654	4,804	−9,146	8,374	10,272	**2,283**
4	−14,510	12,093	−10,326	8,153	−13,822	13,623	14,510	**3,224**
3	−17,240	14,956	−13,887	12,913	−16,944	17,087	17,240	**3,831**
2	−19,234	15,106	−16,611	18,313	−17,258	16,605	19,234	**4,274**
1	−25,706	19,939	−21,783	25,633	−15,900	18,046	25,706	**5,712**

All moments in in.-kip.

Table G21-7   Individual and Peak Member Forces for Analysis Using Amplitude Scaled Records

Item	Ground Motion 1		Ground Motion 2		Ground Motion 3		Peak	Design = Peak × I_e/R
	Min	Max	Min	Max	Min	Max		
Beam 15 m_i	−5,013	7,254	−5,717	5,145	−5,419	5,199	7,254	**1,612**
Beam 15 m_j	−4,998	7,231	−5,701	5,129	−5,405	5,180	7,231	**1,609**
Beam 15 v	−27.8	40.2	−31.7	28.5	−30.1	28.8	40.2	**9.93**
Column 38 m_i	−7,990	5,795	−6,638	6,811	−7,244	6,211	7,990	**1,775**
Column 38 m_j	−7,657	5,272	−5,651	5,942	−6,222	5,095	7,657	**1,701**
Column 38 v	−104.2	73.5	−81.9	84.5	−89.5	65.9	104.0	**23.11**

All moments in in.-kip, shears in kip.

Results Using Spectrum-Matched Ground Motions

Story Displacements and Story Drift

The minimum and maximum *unscaled* interstory drifts computed for each motion are provided in **Table G21-8**. The peak unscaled drift for each level is also provided, as is the maximum design value scaled by C_d/R. The allowable drifts for this Risk Category II building are 2% of the story height. The story drift limit is then $0.02 \times 180 = 3.60$ in. for the 1st and 6th story, and $0.02 \times 150 = 3.0$ in. for stories 2, 3, 4, and 5. The computed drifts are lower than the allowable drifts for all stories.

As may be seen from **Tables G21-8** and **G21-4**, the peak story drifts obtained from the spectrum-matched records are significantly less than for the amplitude-scaled records. This is especially true at the upper stories of the building. The differences are due to the more consistent match between the spectra for the spectrum-matched ground motions relative to

Table G21-8   Individual and Peak Interstory Drift for Analysis Using Spectrum Matched Records

Story	Ground Motion 1		Ground Motion 2		Ground Motion 3		Peak	Design = Peak × C_d/R
	Min	Max	Min	Max	Min	Max		
6	−1.599	2.018	−2.037	1.694	−1.812	1.954	2.037	**1.811**
5	−1.009	1.398	−1.499	1.266	−1.424	1.593	1.593	**1.416**
4	−1.094	0.884	−0.958	1.008	−1.010	1.150	1.150	**1.022**
3	−1.093	0.873	−1.005	0.987	01.094	1.170	1.170	**1.040**
2	−0.871	0.585	−0.839	0.747	−0.944	0.766	0.944	**0.839**
1	−0.986	0.665	−0.883	0.834	−0.994	0.723	0.994	**0.884**

All displacement in in.

the amplitude-scaled motions. For the spectrum-matched motions, the higher-mode influence is significantly reduced due to the better matching at the low periods.

Story Shears, Overturning Moments, and Member Forces

The story shears were computed from the inertial forces, with the results summarized in **Table G21-5**. First, the minimum and maximum unscaled shears are shown at each story for each individual ground motion, and then the peak shears are taken as the maximum magnitude among the three ground motions for each story. Design values are determined by multiplying the peak shears by I_e/R. The values shown for the first story are the same as the base shear.

As seen in **Table G21-9**, the peak design base shear of 132.0 kip is significantly less than that obtained for the amplitude-scaled motions, where the base shear was 161.4 kip. The shear of 132.0 kip is greater than the minimum ELF base shear of 86.3 kip, so additional scaling of force results is not required. It is also noted that the record-to-record variation in story shears is significantly less for the spectrum-matched records than for the amplitude-scaled records. This is likely because the spectra from the three matched records are virtually identical. Similar trends are indicated in the story overturning moments (**Table G21-10**) and member forces (**Table G21-11**).

Comparison of ELF, MRS, and MRH Results

The results from ELF and modal response spectrum (MRS) analyses from Example 20 and the results from the response history analysis for both amplitude-scaled and spectrum-matched ground motions are provided in **Tables G21-12, G21-13,** and **G21-14**. Before comparing the results note that the force results from the MRS analysis were scaled up by a factor of 1.33 to produce a peak base shear equal to 0.85 times the ELF base shear. The

Table G21-9 Individual and Peak Story Shears for Analysis Using Spectrum Matched Records

Story	Ground Motion 1		Ground Motion 2		Ground Motion 3		Peak	Design = Peak × I_e/R
	Min	Max	Min	Max	Min	Max		
6	−143.0	176.0	−175.2	146.5	−150.0	158.8	176.0	**39.1**
5	−146.3	204.2	−214.5	185.7	−208.8	233.7	233.7	**51.9**
4	−320.9	255.7	−253.7	267.0	−273.3	307.7	320.9	**71.3**
3	−364.7	300.2	−324.1	316.8	−357.6	387.8	387.8	**86.2**
2	−464.9	307.0	−445.7	403.4	−503.0	404.1	503.0	**111.8**
1	−594.2	411.1	519.5	500.3	−583.8	427.6	594.2	**132.0**

All shears in kip.

Table G21-10   Individual and Peak Story Overturning Moments for Analysis Using Spectrum-Matched Records

	Ground Motion 1		Ground Motion 2		Ground Motion 3			Design =
Level	Min	Max	Min	Max	Min	Max	Peak	Peak $\times I_e$/R
6	−2,145	2,639	−2,628	2,197	−2,250	2,383	2,639	**586.4**
5	−3,786	5,190	−5,303	4,423	−4,843	5,295	5,303	**1,179**
4	−6,381	6,786	−7,728	7,054	−7,660	8,671	8,671	**1,296**
3	−10,884	9,504	−10,332	11,007	−10,991	12,560	12,560	**2,791**
2	−15,249	13,089	−15,559	14,565	−15,488	12,086	15,488	**3,442**
1	−22,974	17,669	−22,391	21,604	−24,032	22,576	24,032	**5,340**

All moments in in.-kip.

Table G21-11   Individual and Peak Member Forces for Analysis Using Spectrum Matched Records

	Ground Motion 1		Ground Motion 2		Ground Motion 3			Design =
Item	Min	Max	Min	Max	Min	Max	Peak	Peak $\times I_e$/R
Beam 15 m_i	−3,948	5,237	−4,791	5,224	−5,448	5,869	5,869	**1,304**
Beam 15 m_j	−3,936	5,272	−4,777	5,230	−5,431	5,853	5,853	**1,301**
Beam 15 m_j	−21.9	29.0	−26.6	29.1	−30.2	32.6	32.6	**7.24**
Column 38 m_i	−6,576	5,408	−5,613	5,682	−5,957	6,867	6,867	**1,526**
Column 38 m_j	−5,516	4,508	−5,145	4,889	−5,736	5,921	5,921	**1,315**
Column 38 v	−80.4	65.6	−71.3	70.4	−78.0	85.3	85.3	**19.0**

All moments in in.-kip, shears in kip.

Table G21-12   Comparison of Design Interstory Drift

Story	ELF	MRS	MRH (Amplitude-Scaled)	MRH (Matched)
6	1.75	1.51	3.45	1.81
5	1.61	1.20	2.66	1.42
4	1.38	0.93	1.32	1.02
3	1.39	0.94	1.24	1.04
2	0.98	0.71	0.98	0.84
1	0.94	0.76	1.06	0.88

All displacement in in.

force results from the response history analysis did not need additional scaling because the computed base shears were greater than the minimum ELF base shear.

Although significant variation exists among the different sets of results, the variation is not of great concern given all the uncertainties and approximations involved in all the analysis. However, it is somewhat concerning that the results from the spectrum-matched MRH analysis are so low in

Table G21-13   Comparisons of Design Story Shear

Story	ELF	MRS	MRH (Amplitude-Scaled)	MRH (Matched)
6	39.2	42.3	73.1	39.1
5	69.6	58.7	95.1	51.9
4	112.0	84.0	93.7	71.3
3	141.2	103.7	108.2	86.2
2	165.9	125.8	131.9	118.1
1	176.3	150.0	161.4	132.0

All shears in kip.

Table G21-14   Comparisons of Design Member Forces

Level	ELF	MRS	MRH (Amplitude-Scaled)	MRH (Matched)
Beam 15 m_i	2,126	1,529	1,612	1,304
Beam 15 m_j	2,120	1,524	1,609	1,301
Beam 15 m_j	11.9	8.48	9.93	7.24
Column 38 m_i	2,494	1,836	1,775	1,526
Column 38 m_j	2,195	1,596	1,701	1,315
Column 38 v	31.3	23.1	23.1	19.0

All moments in in.-kip, all shears in kip.

comparison with ELF and MRS, particularly with regard to base shear and member force. The different results could be eliminated by requiring that the force results from the MRS analysis and MRH analysis be scaled in a consistent manner, e.g., to 85% of the ELF base shear, regardless of the controlling value of C_s.

This and other inconsistencies between the response linear history provisions in Chapter 16 and the modal response spectrum procedures in Chapter 12 lead to the conclusion that the linear modal response history analysis procedure should be moved to Chapter 12 of ASCE 7 and be made as consistent as possible with the response spectrum approach. In addition to the force scaling requirements already mentioned, amplitude scaling of ground motions should be performed over the period range of $0.2T$ to $1.0T$, instead of $0.2T$ to $1.5T$. Additionally the code should specify how many modes need be used in a modal response history analysis. Finally, spectrum matching should be specifically allowed for all linear response history analysis. In the future, matched ground motions could be provided by a simple web-based utility.

Endnote

[1] Section 16.1.4.1 of Supplement 1 of ASCE 7-10 requires scaling requirements for linear response history analysis to be consistent with those of the modal response spectrum method. This update is not included in the example.

Example 22

Diaphragm Forces

This example discusses Section 12.10 of ASCE 7, which covers the computation of in-plane floor diaphragm forces, including collector and chord elements. Diaphragm forces caused by both inertial and system effects are included.

The structure considered in this example is a six-story office building located near Memphis, Tennessee. The pertinent information for the building and the building site are as follows:

> Site Class = B,
> $S_S = 0.6\,g$,
> $S_1 = 0.2\,g$,
> $F_a = 1.0$ (from Table 11.4-1),
> $F_v = 1.0$ (from Table 11.4-2),
> $S_{DS} = (2/3)\,S_S \times F_a = 0.400\,g$ [Eqs. (11.4-1) and (11.4-3)],
> $S_{D1} = (2/3)\,S_1 \times F_v = 0.133\,g$ [Eqs. (11.4-2) and (11.4-4)],
> Risk Category = II (from Table 1-1),
> Importance factor $I_e = 1.0$ (from Table 11.5-1), and
> Seismic Design Category = C (from Tables 11.6-1 and 11.6-2).

A plan view of the structural system for the building is shown in **Fig. G22-1**. Intermediate steel moment frames resist all forces in the east–west direction and a dual intermediate moment frame–special concentrically braced frame system is used in the north–south direction. The lowest story has a height of 15 ft, and the upper stories each have a height of 12.5 ft. The total seismic weight of the system W is 7,500 kip.

This example considers loads acting in the north–south direction only. The design values for the dual system are determined from Table 12.2-1 and are summarized as follows:

> $R = 6$,
> $\Omega_o = 2.5$,
> $C_d = 5.0$, and
> Height limit = None.

Fig. G22-1

Typical floor plan of a six-story building

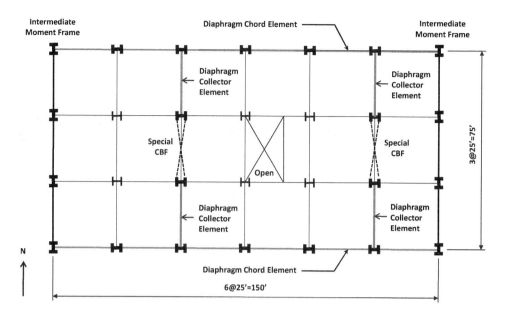

The period of vibration from structural analysis is 0.73 s. This period controls over the upper limit period $T = C_u T_a$, which is 0.85 s for this structure.

The seismic response coefficient is taken as the larger of the values computed from Eqs. (12.8-3) and (12.8-5):

$$C_s = \frac{S_{D1}}{T(R/I_e)} = \frac{0.133}{0.73(6/1)} = 0.0304$$

$$C_s = 0.044 S_{DS} I_e = 0.044(0.40)(1) = 0.0176 > 0.01$$

The design base shear (Eq. 12.8-1) is

$$V = C_s W = 0.0304(7,500) = 228 \text{ kip}$$

The floor deck consists of a 4.5-in. concrete slab over metal deck. This slab must resist both inertial forces and forces developed because of shear transfer between the lateral force–resisting elements. Also considered in the analysis are the collector elements (drag struts) and diaphragm chords. These elements are shown in **Fig. G22-1**. The elements shown as diaphragm chord elements act as collector elements when seismic forces act in the east–west direction.

The inertial forces at a given level x are computed in accordance with Section 12.10.1.1 and Eq. (12.10-1):

$$F_{px} = \frac{\sum_{i=x}^{n} F_i}{\sum_{i=x}^{n} w_i} w_{px} = q_{px} w_{px} \qquad \text{(Eq. 12.10-1)}$$

where F_i is the lateral force applied to level i, w_i is the weight at level i, w_{px} is the weight of the diaphragm (or portion thereof) at level x, and n is the

Table G22-1 Diaphragm Force Coefficients for a Six-Story Building

Level	w (kip)	F (kip)	Σw (kip)	ΣF (kip)	$q = \Sigma F/\Sigma w$	$q_{min} = 0.2S_{DS}I_e$	$q_{max} = 0.4S_{DS}I_e$	Controlling F_{px} (kip)
6	1,150	62.5	1,150	62.5	0.054	0.080	0.160	92.0
5	1,250	55.8	2,400	118.3	0.049	0.080	0.160	100.0
4	1,250	44.0	3,650	162.3	0.044	0.080	0.160	100.0
3	1,250	32.5	4,900	194.8	0.040	0.080	0.160	100.0
2	1,250	21.4	6,150	216.2	0.035	0.080	0.160	100.0
1	1,350	11.8	7,500	228.0	0.030	0.080	0.160	108.0

number of levels. The term q_{px} is not explicitly used in ASCE 7, but is a convenient parameter to determine. The force F_{px} that is obtained from Eq. (12.10-1) shall not be less than $0.2S_{DS}I_e w_{px}$ but need not exceed $0.4S_{DS}I_e w_{px}$. The results of the analysis for the coefficient q is given in **Table G22-1**. The minimum value given by $0.2S_{DS}I_e$ controls at each level.

On the basis of the previous calculations, each level of the diaphragm must be designed for a force of 0.08 times the seismic weight at the level of interest. For the fifth level, for example, the total diaphragm force is $0.08(1,250) = 100.0$ kip. From a structural analysis perspective, the most accurate way to determine the stresses and forces in the various components of the diaphragm would be to model the diaphragm with shell elements and distribute the 100-kip force to the individual nodes of the element on a "tributary mass" basis. Thus, the diaphragm would be modeled as semirigid, even though Section 12.3.1.2 would define this diaphragm as rigid. Other analysis approaches are also available, including the linear collector method, the distributed collector method, and the strut and tie modeling method. See Sabelli et al. (2009) for a description of these methods.

Additional comments regarding the analysis and design of the diaphragm components are as follows:

1. Discontinuities in the lateral load–resisting elements above and below the diaphragm cause forces to be transferred between the lateral load–resisting elements and impose in-plane forces in the diaphragm. These forces must be added to the forces developed from the diaphragm forces F_{px} that were determined with Eq. (12.10-1). In accordance with Section 12.12.1.1, elements resisting the transfer forces must be designed with the redundancy factor ρ that has been determined for the structural system, but the forces caused by the application of F_{px} alone may be designed with a redundancy factor of 1.0.

2. In SDC C and above, collector elements must be designed using the overstrength factor Ω_o that has been assigned to the structural system resisting forces in the direction of F_{px}.

3. Special care has to be taken to produce realistic diaphragm, collector, and chord forces when the diaphragm is modeled as rigid in a three-dimensional structural analysis.

4. Recovering diaphragm forces from a modal response spectrum (MRS) analysis is not straightforward. Using the approach outlined in Section 12.10 is reasonable for determining diaphragm forces, even when MRS has been used for the analysis of the lateral load–resisting system. However, as mentioned in point 1, the total diaphragm forces must include the transfer forces, if present, and these forces must be recovered from the MRS analysis.

Frequently Asked Questions

The following list of "Frequently Asked Questions" was provided by practicing engineers that use ASCE 7 on a regular basis. Some of the issues in these questions are addressed in the main body of the *Guide*, while others are not. With few exceptions, the answers provided in the following are given without reference to the material in the *Guide*, thereby allowing this FAQ section to stand alone.

1. Are there any specific guidelines to meet the provisions of Section 1.4: General Structural Integrity?

No specific requirements currently exist for achieving structural integrity in ASCE 7. However, seismic detailing such as that required in Seismic Design Categories C and above will provide continuity, redundancy, and ductility.

The 2012 IBC introduces requirements associated with structural integrity in Section 1614. These requirements are only specified for a limited number of cases, but could be used, voluntarily, for any building. It's not certain that the provisions of Section 1.4 will be met by following Section 1614, but the overall integrity of the building will be improved.

2. Does the interconnection requirement of Section 12.1.3 apply across any section cut a designer might draw across a diaphragm?

Section 12.1.3 provides requirements for a continuous load path and interconnection of all parts of the structure. These requirements would apply across any section cut through a diaphragm. Additionally, the diaphragm must be analyzed and designed in accordance with Section 12.10.

3. How are story stiffnesses calculated when determining if a vertical structural irregularity of Type 1a or 1b exists?

Table 12.3-2 discusses the conditions under which a vertical structural irregularity exists. Stiffness irregularities (Types 1a and 1b) occur when the lateral stiffness of one story is less than a certain percentage of the lateral stiffness of the story above it (or less than a certain percentage of the average stiffness of the three stories above it). Note, however, that it is only necessary to determine if such irregularities exist for buildings in SDC D and above because stiffness irregularities in buildings assigned to SDC C and lower have no consequences.

To address the issue of determining story stiffness, consider the structure shown in **Fig. FAQ-1**. This structure is a one-bay moment-resisting frame. The column stiffness is the same at each level, and the beam stiffness is the same at each level. Three different configurations were analyzed, in which each system had a different beam-to-column stiffness ratio as follows,

Beam-to-column stiffness ratio = 0.01: Structure behaves like a cantilever beam,

Beam-to-column stiffness ratio = 1.00: Structure behaves like a moment frame, and

Beam-to-column stiffness ratio = 100: Structure behaves like a shear frame.

Fig. FAQ-1
System and loading to determine story stiffness

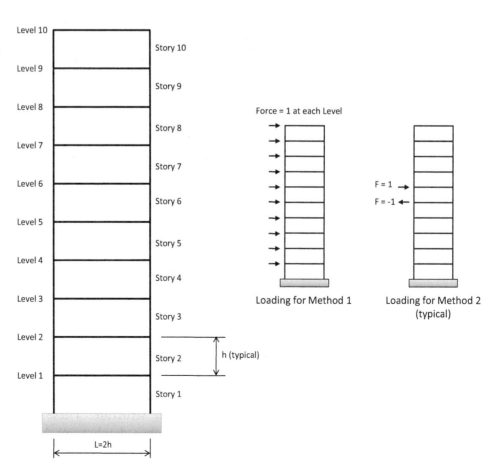

Note that for all systems the columns are fixed at the base. Two different methods were used to determine the lateral stiffness of the story:

Method 1: Load the structure with a unit lateral force at each level (all levels loaded simultaneously), compute the interstory drift at each level, and determine the stiffness at each story as the story shear divided by the story drift. Only one loading is required for this method.

Method 2: Load the structure with a positive unit force at level i and a negative unit force at level i-1, compute the interstory drift between the two loaded levels, and invert to obtain the stiffness of the story between levels i and i-1. Ten separate loadings are required for this method.

The results of the analysis are shown in **Fig. FAQ-2**. Part a of this figure shows results from the system with a beam-to-column stiffness ratio of 0.01. Note that the two methods produce dramatically different story stiffnesses and that for both methods the stiffness increases significantly at the lowest level. This is particularly true when Method 2 is used. This increase in stiffness at the lowest level is due to the fixed support at the base. The change in stiffness along the height is due to the support condition and to rigid body rotation in the upper floors and not to any actual variation in the stiffness of the structural system.

The results for the system with the beam-to-column stiffness ratio of 1.0 are presented in **Fig. FAQ-2(b)**. As before, the results are dramatically different for the two methods of analysis, with Method 2 providing the larger story stiffness. For Method 1 stiffness varies significantly along the height, but for Method 2 stiffness remains somewhat uniform, except for the first level. The increase in stiffness at the first level is due to the fixed base condition.

In **Fig. FAQ-2(c)**, the two methods give very similar results, with Method 2 producing a consistently greater stiffness than Method 1. While the fixed base condition has some influence on the results, this is less significant that it was for the other two beam-to-column stiffness ratios.

Method 2 appears to produce better results than Method 1 because the story stiffnesses reported by Method 2 are more uniform (as would be expected for a structure with uniform properties along the height). The accumulated rotations of the stories below the story of interest strongly influence the results of Method 1. For example, for the system with the lowest beam-to-column stiffness ratio, almost all the drift in the upper levels is due to rigid body rotation.

The basic conclusions from the analysis are as follows:

1. Method 1 should not be used to determine story stiffness.
2. Method 2 is preferred, but the computed stiffnesses at the lower levels may be artificially high due to the fixed boundary conditions, and the stiffness at the upper levels may be artificially low due to the presence of rigid body rotations. The method appears to be reliable for moment-resisting frames, but may produce unrealistic estimates of story stiffness in systems that deform like a cantilever (e.g. tall slender shearwalls and braced frames).

Fig. FAQ-2

Results of stiffness
analysis of a 10-story
building system

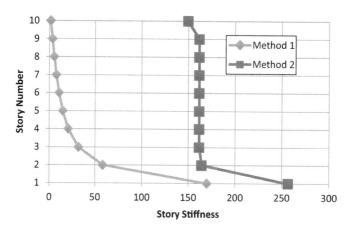

(a) Beam-to-column stiffness ratio = 0.01

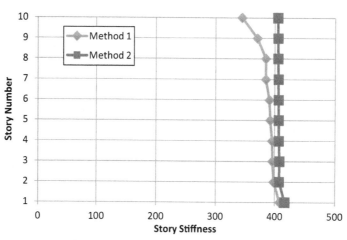

(b) Beam-to-column stiffness ratio = 1.0

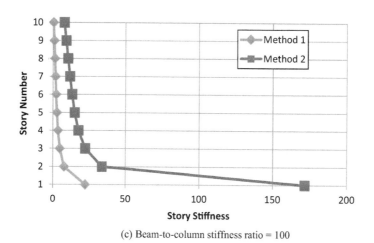

(c) Beam-to-column stiffness ratio = 100

4. How are story strengths calculated when determining if a vertical structural irregularity of Type 5a or 5b as in Table 12.3-2 exists?

Calculating the strength of a "story" of a lateral load–resisting seismic
system is difficult, if not impossible. The computed strength depends on the
loading pattern, the location of yielding throughout the story, and on the

Fig. FAQ-3

Three "mechanisms" for computing story strength

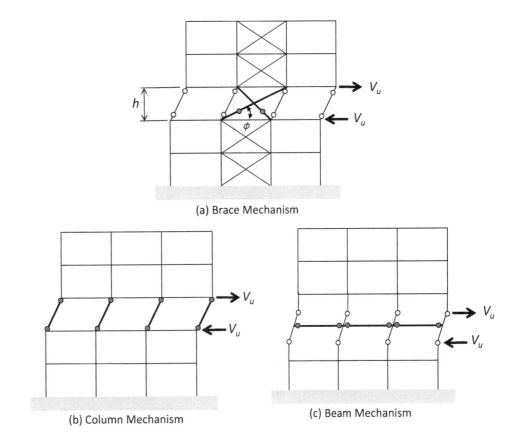

(a) Brace Mechanism

(b) Column Mechanism

(c) Beam Mechanism

capacities of the yielding elements. The element capacities are a function of the materials used, the details of the cross section, and the forces that act on the section. For example, the flexural strength of a reinforced concrete shear wall is a function of the axial compressive force in the wall. Similarly, the flexural capacity of a steel or concrete column is a function of the axial force in the column. Shear capacities of concrete sections are also a function of the axial force in the section.

Estimating the story capacities of some simple systems, such as those shown in **Fig. FAQ-3**, is possible. For system a, which is a braced frame system, the story capacity can be based on the strength of the braces and would be as follows:

$$V_u = (F_{uC} + F_{uT})\cos\phi \qquad \text{(Eq. FAQ-1)}$$

where F_{uC} and F_{uT} are the compressive and tensile capacities of the braces, respectively, and ϕ is the angle shown in the figure. This capacity assumes that the columns do not yield axially and have a moment release (moment-free hinge) at the top and bottom of the story. If the columns are also assumed to yield, the strength from a column mechanism (see **Fig. FAQ-3(b)** and Eq. [FAQ-2]) may be added to the strength obtained from Eq. (FAQ-1).

For a moment-resisting system, such as shown in **Fig. FAQ-3(b)**, the story capacity may be based on a sway mechanism. This method is based on the assumption that plastic hinges from in the top and bottom of each column of a particular story. If the flexural capacities of the columns are known, the story strength may be obtained as

$$V_u = \frac{2}{h} \sum_{i=1}^{ncols} M_{uC,i}$$ (Eq. FAQ-2)

where h is the story height, *ncols* is the number of columns in the story, and M_{uC} is the flexural capacity of the column hinges at the top and bottom of the columns (which may be a function of the axial force in the column). This type of mechanism might form in columns of ordinary and intermediate moment frames, but it is unlikely to occur in special moment frames because of strong column–weak beam design requirements.

A second type of story capacity may be computed on the basis of the beam strengths. The mechanism for computing the story capacity is shown in **Fig. FAQ-3(c)**. This computed capacity is

$$V_u = \frac{1}{h} \sum_{i=1}^{nbays} \left(M_{uB,i}^+ + M_{uB,i}^- \right)$$ (Eq. FAQ-3)

where *nbays* is the number of bays, M_{uB}^+ is the positive moment flexural capacity at one end of the beam, M_{uB}^- is the negative moment capacity at the other end of the beam, and *nbays* is the number of bays. It is noted, however, that a mechanism consisting of plastic hinges at each end of each beam in a *single story* is impossible (without loss of continuity in the columns above and below the level in question). Note that equations similar to Eqs. (FAQ-2) and (FAQ-3) are discussed in Part C3 of the commentary on the *Seismic Provisions for Structural Steel Buildings* (AISC, 2005).

There is no straightforward way to compute the story shear capacity of a shear wall. Determination of the story capacity of combined systems and dual systems is also problematic.

ASCE 7 has only two consequences when weak-story irregularities occur. The first of these is given in Section 12.3.3.1, which prohibits structures in SDC E and F from having a Type 5a or 5b vertical irregularity and structures in SDC D from having a Type 5b irregularity. The second consequence is given in Section 12.3.3.2, which states that buildings with vertical irregularity Type 5b must be limited to 30 ft in height (with certain exceptions). Note also that weak-story irregularities do not prohibit the use of the equivalent lateral force method of analysis, whereas soft story irregularities in SDC D and above buildings may prohibit the use of ELF.

Aside from determining if weak story irregularity exists, story shear capacity may also be needed in association with computing the redundancy factor (Section 12.3.4) and for determining the limiting value of the stability coefficient (Eq. [12.8-17]).

5. Why are forces for columns that support discontinuous braced frames amplified by the overstrength coefficient Ω_0, but not columns in braced frames that are continuous?

This requirement, from Section 12.3.3.3, applies only to structures with an in-plane, Type 4 horizontal irregularity or an out-of-plane offset, Type 4 vertical irregularity in the lateral load–resisting system. Experience from previous earthquakes has indicated that such irregularities impose extreme demands on the portion of the structure below the irregularity, and such irregularities have been identified as a significant contributor to the partial or complete collapse of structures during earthquakes. The amplification factor serves as a "penalty," discouraging the use of such irregularities and reducing the likelihood of severe damage or collapse if such irregularities exist.

6. How is the redundancy factor ρ calculated for walls with *h/w* < 1.0?

Section 12.3.4.2 states that the redundancy factor ρ must be taken as 1.3 for buildings in SDC D and above unless one or both of two conditions are met. One of these conditions is that "each story resisting more than 35% of the base shear in the direction of interest shall comply with Table 12.3-3." The intent of Table 12.3-3 is that the engineer consider each lateral load–resisting element in each direction and perform the test associated with that element. For example, consider a system with two moment frames (A and B) and one braced frame (C) resisting loads in a given direction. Analysis would be performed with moment releases placed at each end of a given beam in moment frame A, with frames B and C intact. If the placement of the releases does not reduce the system strength by more than 33%, or cause an extreme torsional irregularity, the redundancy factor can be taken as 1.0. Note that theoretically the test must be performed once for each beam in frame A, then again for each beam in frame B, and then again for each diagonal in the braced frame C. Fortunately, the strength check may often be avoided by inspection. However, the extreme torsion irregularity check may not be as easy to visualize.

Note that no requirement exists to remove walls with a height-to-width ratio of less than 1.0. Thus if the structure described above had two moment frames and one wall with the wall having *h/w* less than 1, the wall would never need to be removed in the redundancy analysis. *If the system consisted only of walls, with each wall having* h/w *less than 1, no walls would need to be removed, and the redundancy factor would default to 1.0.* Note that this would even be the case for a system with *only* one or two walls (with *h/w* less than 1) in a given direction. Such systems do not seem to be particularly redundant and should be designed with ρ = 1.3. This system appears to have "fallen through the cracks" in ASCE 7, resulting in a potentially unconservative design.

7. Are P-delta effects calculated based on the initial elastic stiffness, or are they analyzed at the design story drift?

The stability coefficient θ computed by Eq. (12.8-16) has two uses. First, it is used to determine if including P-delta effects in the analysis is necessary. If θ is less than or equal to 0.1 P-delta effects may be neglected, and if θ is greater than 0.1, such effects must be included. Second, when θ > 0.1, θ is used to amplify both the displacements and the member forces, the amplification factor is given by $1/(1 - θ)$.

The stiffness used to calculate displacements in Eq. (12.8-16) should be the same stiffness used to compute the period of vibration of the structure and to compute the design story drifts used in accordance with the allowable story drift of Table 12.12-1. The analytical model used to compute the stiffness should conform to the requirements of Section 12.7.3. These modeling requirements are consistent with a structure subjected to service level loads.

8. Why is the P-delta stability index θ required to be less than 0.5/(βC$_d$) in earthquake engineering? This results in permitted P-delta effects much less than what is allowed in nonseismic design. Considering no failures reported due to P-delta during earthquake, why such a strict criteria that generally controls the design?

The limit on the stability coefficient provides two effects. First, it protects buildings in low seismic hazard regions against the possibility of postearthquake (residual deformation triggered) failure. Second, it provides a limit in the implied overstrength of a building.

Regarding P-delta–triggered failures, P-delta effects are generally more critical in buildings in low and moderate hazard areas (where buildings have relatively low lateral stiffness) than they are in high hazard areas (where the stiffness is relatively high). Limited knowledge exists on the likely performance of code-compliant buildings in the lower hazard areas, but without some limit on the stability ratio, it is entirely possible that failures may occur to dynamic instability.

The term β in the stability coefficient is essentially the inverse of the overstrength of the story. When computed, the overstrength often exceeds 2.0. Thus it may be beneficial to compute the β factor when it appears that the upper limit on θ is controlling. Unfortunately, computing the story overstrength is not straightforward (see FAQ 3).

9. Is checking P-delta effects necessary (per Section 12.8.7), when such effects are automatically included in the structural analysis?

When the P-delta analysis is performed by a computer, the displacements and story shears are automatically amplified, so amplifying these quantities using the ratio $1/(1 − θ)$ is not necessary. However, determining if the maximum allowable value of θ, given by Eq. (12.8-17), has been exceeded is necessary. To do this, recovering the stability coefficient from the structural analysis is required. This coefficient can be recovered by performing the analysis with and without P-delta effects and computing the interstory drifts from each analysis. If the interstory drift from the analysis without P-delta included is designated Δ_o, and the interstory drift from the analysis with P-delta effects included is Δ_f, the story stability coefficient is given by

$$\theta = 1 - \frac{\Delta_0}{\Delta_f}$$

(Eq. FAQ-4)

If the computed value of θ for *each level* is less than 0.1, the analysis may be re-run with P-delta effects turned off. If θ is greater than θ_{max} for any story the structure must be re-proportioned.

10. How much eccentricity in EBF is required for a C_t value of 0.03 and $x = 0.75$ for determination of the approximate period?

Table 12.8-2 provides parameters C_t and x to be used in the determination of the approximate period, T_a. The parameters $C_t = 0.03$ and $x = 0.75$, applicable to eccentrically braced frames, may be used *only* if the frame is designed and detailed as an EBF in accordance with the requirements of the *Seismic Provisions for Structural Steel Buildings* (AISC, 2010b).

Theoretically, a frame with eccentric connections may be used as part of an $R = 3$, "steel system not specifically detailed for seismic resistance" (System Type H in Table 12.2-1). However, in this circumstance nothing prevents the designer from using, say, a 6 in. eccentricity, which would result in a frame stiffness closer to that of a concentrically braced frame than that of an eccentrically braced frame. For this reason, the lack of limits on the geometry of $R = 3$ EBFs, the coefficients in Table 12.8-2 are limited to systems specifically designed as EBFs.

11. Section 12.7.3 has requirements for structural modeling and requires that cracked section behavior be considered for concrete structures. Specifically, how is concrete cracking included in the analytical model?

The cracked properties of reinforced concrete are difficult to determine because they depend on the type (axial, bending, or shear) of action and the magnitude of the action. Expressions for determining cracked section properties may be found in several textbooks, including Park and Paulay (1974) and Paulay and Priestly (1992).

Table 6-5 of ASCE 41-06 (ASCE, 2007) provides estimates of cracked section properties for beams, columns, walls, and slabs. Cracked properties are provided for sections under flexure, shear, and axial load. Flexural properties are as low as 50% of the gross property (e.g. for nonprestressed beams), but the axial and shear rigidities are not reduced to account for cracking. For axial, this is appropriate if the member is in compression. If for any reason the member is in net tension and the concrete stress exceeds the cracking stress (a rare condition), the axial rigidity should be reduced accordingly. For shear, ASCE 7 establishes a rigidity of $0.4E_cA_g$, where E_c is Young's modulus of the concrete, and A_g is the gross area. It is very important to note here that the 0.4 factor is *not* a reduction factor for cracking. Instead, it is the factor that converts Young's modulus (E) to the shear modulus (G). Analysts should carefully consider the effect of shear cracking, as the reduction in section rigidity to shear cracking can exceed the reduction due to flexure. This is particularly true for squat shear walls for link beams in coupled wall construction where the length-to-depth ratio of the coupling beams is less than about 4.

12. Section 12.7.3 has requirements for structural modeling and requires that deformations in the panel zone of steel frames be considered. Specifically, how are such deformations included in the analytical model?

Shear deformations in the panel zones of beam-column joints can be an important source of drift in steel moment frames and, for some systems, can be responsible for as much as 40% of the total drift (Charney, 1990). The most effective way to include shear deformations is to explicitly model the beam-column joint using an assemblage of rigid links and rotational springs (Charney and Marshall, 2006). Many modern computer programs provide this type of model. When such a model is not used, the next-best approach is to model the frame using centerline dimensions, meaning that the flexible length of the beam is equal to the span length between column *centerlines*. This type of model will overestimate flexural deformations in the panel zone and underestimate shear deformations in the panel zone. The errors are offsetting, so the net result is a reasonably accurate analysis. Note, however, that in some cases the centerline model is unconservative, meaning it will produce results that underpredict the deflections.

Some programs allow the use of a rigid end zone, but this should not be used unless the panel zone is reinforced with doubler plates, and even in this case, only a portion of the panel zone dimensions should be considered rigid.

13. What is the approximate period of a dual system?

Table 12.8-2 does not include dual systems (or other combined systems), so such systems automatically default to "other structural systems," wherein the values for parameters C_t and x are 0.02 and 0.75, respectively. These parameters seem to produce a somewhat low period for a dual system, but given that the stiffness of such systems is likely dominated by the stiffer component (e.g., the shear wall in a frame-wall system), the degree of conservatism is probably not excessive.

Also of concern is the determination of periods for combined systems such as those shown as Buildings C and D in **Fig. G8-2**. In such situations the default parameters may be excessively conservative, but perhaps this is warranted as a penalty for using nontraditional (and not well understood) structural systems.

If the periods for dual systems and combined systems are determined analytically (using a computer), the degree of conservatism associated with the use of the default parameters may be reduced by using the upper limit period, $C_u T_a$, when appropriate.

14. Should drifts be calculated at the center of mass or at the diaphragm corners for comparison to drift limits?

According to Section 12.8.6, drifts are defined as the difference between the *displacements at the centers of mass* of adjacent stories. However, if the structure is assigned to Seismic Design Category C or higher and has a torsion irregularity or an extreme torsion irregularity, Section 12.12.1 requires that the drifts be computed as the difference between *displacements at the edge of the story.*

15. Table 12-2-1 provides design coefficients for cantilever column systems. The various coefficients (R, C_d, Ω_o, height limit, etc.) depend on the type of detailing used in the cantilever column system. All systems in this section are frames that are made up of columns and beam. How are frame detailing requirements applied to cantilever columns?

The various material specifications provide several requirements for the framing systems that are designated in part G of Table 12.2-1. For example, for columns in special reinforced concrete moment frames, ACI 318 limits material properties, cross sectional dimensions, area of reinforcement, spacing of reinforcement, and location of splices. Additionally, detailing requirements are provided for detailing and spacing of transverse reinforcement. These limitations and requirements would be applicable to cantilever systems with a "special reinforced concrete moment frame" as the designated seismic force–resisting system. Certain rules, such as strong column–weak beam requirements are clearly not applicable to cantilever systems (which do not have beams).

Note that Section 11.2 defines cantilever systems as "a seismic force–resisting system in which lateral forces are resisted entirely by columns acting as cantilevers from the base." Note also that Section 12.2.5.2 limits the axial load that the cantilever system can carry and requires that the foundation for such systems be designed with the applicable overstrength factor Ω_0, which is 1.25 for all cantilever systems constructed from steel or concrete. Finally, note that cantilever systems are subject to the redundancy requirements of Section 12.3.4.2.

16. I am designing a pedestrian bridge support (hammer head) where in the longitudinal direction, I can have frame action but in the transverse direction, I have to cantilever the support. If the height of the structure is more than 35 ft no system in Table 12.2-1 under "cantilever column system" would be allowed. Can I designate the support described above as shear wall and use the R, C_d, Ω_o, and height limit provided for shear wall? More importantly, why is there a difference in cantilevered system and shear wall system?

If the support could be designed and detailed in accordance with all requirements for a wall, it would seem that a wall system could be used in the transverse direction. However, an element with a length-to-thickness ratio less than, say, 4, is unlikely to be able to be detailed to meet all the requirements of a wall. Note that for special concrete moment frames, columns with a length-to-thickness ratio greater than 2.5 (1/0.4) are not allowed and must thereby be designated as walls.

The lower design values and height limitations for cantilever systems are due to the generally poor performance of these systems (compared with walls). The poor performance relates to low lateral stiffness and a lack of redundancy of cantilever systems.

17. Section 11.8.3, which applies for SDC D, E, and F, requires that the geotechnical report include "lateral pressures on basement and retaining walls due to earthquake motions." Does that mean these pressures can be taken as zero in SDC B and C?

Lateral pressures due to ground shaking may be neglected in buildings assigned to SDC B and C. Forces and stresses induced by such pressures in buildings assigned to SDC D through F would be considered as part of the earthquake load E in the pertinent load combinations of Chapter 2.

18. Where do the R, C_d, and Ω_o values in Table 12.2-1 come from?

All values currently in Table 12.2-1 are completely qualitative. They are based on engineering judgment and on experience gained from building behavior in past earthquakes. R and C_d values first appeared in the ATC 3-06 report (ATC, 1984), and only 21 sets of values were provided. Over the years additional systems were added, or existing systems were subdivided into several systems. ASCE 7-10 includes 85 sets of values. The term Ω_o first appeared in ASCE 7-98.

More recently, an analytical procedure was developed to provide the R, C_d, and Ω_o values. This procedure, described in FEMA P-695 (FEMA, 2009b) states that for a given class of system (such as a special steel moment frame), R should be set such that no more than a 10% probability exists that the building will collapse when subjected to the maximum considered earthquake (MCE) ground motion. In the procedure, various "archetypes" of the system are designed assuming some trial R value, and each archetype is analyzed, using nonlinear dynamic response history procedures, for 44 preselected ground motions. The analysis is necessarily quite detailed to capture all anticipated nonlinear effects (yielding, P-delta effects). For each archetype and for each ground motion the ground motion's intensity is increased until the system collapses. The ground motion intensity at which half of the earthquakes cause collapse of the archetype is used to compute the collapse margin ratio for that archetype. Using that value and various adjustments, including those for uncertainty, the probability of collapse can be computed. If no individual archetype has more than a 20% probability of collapse and if the average probability of collapse for all archetypes is less than 10%, the system passes and the R value used for design is deemed appropriate. If the probabilities of collapse are excessive the procedure is repeated with a smaller R value, and if the probabilities of collapse are too low (indicating an overly conservative design) the process is repeated with a lower R value. The R value is continually adjusted until the desired behavior is obtained. At that point C_d is taken as equal to R. Ω_o is determined separately from a nonlinear static pushover analysis.

The procedure as described is extremely time consuming due to the necessity to design several archetypes and run thousands of nonlinear response history analyses. NIST (2010b) provides some example applications of the procedure. It is anticipated that some of the new systems proposed for ASCE 7-16 will have design values based on the FEMA P-695 procedure. To date, none of the existing structural systems (those currently listed in Table 12.2-1 of ASCE 7-10) have been completely vetted by the procedure, and it is not anticipated that such a vetting will occur prior to the publication of ASCE 7-16.

19. Doesn't ASCE 7 contain an inherent contradiction in that the effective seismic weight used in computing equivalent lateral forces is based on 25% of live load (LL) (for storage loads), while the gravity effects include at least 0.5LL (up to 1.0LL) for strength design?

First, it should be noted that Section 12.7.2 indicates that the 25% LL requirement is the *minimum* for storage live loads. It is feasible, if not likely, that higher loads would be used. The reason that a lower LL is used for effective seismic weight than for gravity is that the loads are (probably) not rigidly attached to the structure and would therefore not be subjected to the same lateral accelerations (and inertial forces) as the structure. Additionally, the sliding movement of the storage load dissipates energy, increasing the effective damping in the system. However, the gravity live load is constant, even when the building shakes laterally. The same "inconsistency" occurs with partition loads and snow loads, wherein the amount of load included in the effective seismic weight is less than that used for gravity load effects.

20. Why was 65 ft chosen as the height limit in Section 12.2.5.6?

There is no specific reason other than the fact that 65 ft (five to six stories) is a reasonable delineator between low-rise and high-rise buildings.

21. How are diaphragm forces (shears and collector and chord forces) calculated when using modal response spectrum analysis?

Section cuts above and below the floors can be used to determine the diaphragm forces associated with the vertical distribution of seismic forces determined from the MRS analysis. These forces would be applied in a static manner, similar to the forces computed using Eq. (12.10-1). Note that the minimum diaphragm force requirements of Section 12.10 still apply when MRS analysis is used.

A more accurate distribution of diaphragm forces may be achieved from MRS analysis if the diaphragms were physically modeled using shell elements. In this case, the analyst must be able to define a cut through the diaphragm for which net forces (e.g., shear through the cut) are determined for each mode and then combined using SRSS or CQC (see Section 12.9.3).

22. How is the overstrength factor Ω_o used? Are elements designed with the factor expected to remain elastic during an earthquake?

The overstrength factor is used in load combinations 5 and 7 in strength design and in load combinations 5, 6, and 7 in allowable stress design

(Section 12.4.3.2). This factor is applied only to certain elements, *and never to the structure as a whole.* Specific cases in ASCE 7 where the overstrength factor is used include

1. Design of elements supporting discontinuous wall or frames (Section 12.3.3.3), which applies only to systems with horizontal irregularity Type 4 of Table 12.3-1, or vertical irregularity Type 4 of Table 12.3-2;
2. Design of collector elements in SDC C, D, E, or F (Section 12.10.2.1); and
3. Foundations of cantilever columns systems (Section 12.2.5.2).

Various material specifications may also require design with the overstrength factor. For example, Section 8.3 of the *Seismic Provisions for Structural Steel Buildings* (AISC, 2010b) requires that in the absence of applied moment the required axial and tensile column strength shall be determined on the basis of the amplified seismic load, where the amplified seismic load is defined as that load combination that includes the overstrength factor Ω_o. Numerous other cases must consider the overstrength factor in steel design. ACI 318-05 also refers to the overstrength factor. For example, the factor is used in association with the design of elements of concrete diaphragms (see Section 21.9.5.3 of ACI 318).

Elements designed with the overstrength factor will not neccesarily remain elastic during an earthquake. However, these elements are expected to suffer less damage than elements not designed with the overstrength factor and have a lower likelihood of failure.

23. What is the purpose of the exponent *k* in Eq. (12.8-12)?

The exponent k accounts, in an approximate manner, for higher mode effects when distributing the design base shear along the height of the structure. For a structure with a uniform story height and story mass, $k = 1$ (for relatively short buildings with T less than or equal to $0.5\,\mathrm{s}$) will produce a straight-line upper-triangular lateral force pattern, and $k = 2$ (for relatively tall buildings for T greater than or equal to $2.5\,\mathrm{s}$) will produce a parabolic force distribution with increasing slope at higher elevations.

Note that the limit on using the ELF method of analysis for SDC D, E, and F buildings with $T > 3.5\,T_s$ (see Table 12.6-1) is based on calculations by Chopra (2011) that show that the ELF method may be unconservative when $T > 3.5\,T_s$.

24. When using the modal response spectrum method of analysis (Section 12.9), I lose all the signs of member forces due to the SRSS or CQC combinations. Sometimes knowing these signs is useful (e.g., is the moment positive or negative?). Is there a way to recover the signs?

In general, recovering the signs is not possible. If maintaining the signs is important, the analyst should consider using the linear response history analysis method described in Chapter 16 of ASCE 7-10. Probably the greatest challenge to using this approach is the selection and scaling of appropriate ground motions. An efficient methodology for performing such analysis (which adheres to the requirements of Chapter 16) is provided by Aswegan and Charney (2014). Key to this approach is the use of spectrum-matched ground motions that are derived from natural ground motions.

In the linear response history method the maximum positive and negative force results are provided for each ground motion, and either the average or the envelope of these forces is used for design. In addition to retaining the signs, concurrent actions may be reported, meaning that the axial force present in a column when it reaches peak moment is available for checking axial-force bending-moment interaction.

Appendix A

Interpolation Functions

ASCE 7 provides tables from which the user must interpolate to determine the necessary values. This appendix provides a graphical representation of the data provided in several of these tables and mathematical functions from which the appropriate values may be obtained without interpolation.

Table GA-1 Interpolation Formulas for Computing Site Class Factor F_a

| Site Class | Value of F_a for a Given Range of Values of S_s | | | | | |
	(1) <0.25	(2) 0.25–0.50	(3) 0.50–0.75	(4) 0.75–1.0	(5) 1.0–1.25	(6) >1.25
A	0.8	0.8	0.8	0.8	0.8	0.8
B	1.0	1.0	1.0	1.0	1.0	1.0
C	1.2	1.2	1.4–$0.4\,S_s$	1.4–$0.4\,S_s$	1.0	1.0
D	1.6	1.8–$0.8\,S_s$	1.8–$0.8\,S_s$	1.5–$0.4\,S_s$	1.5–$0.4\,S_s$	1.0
E	2.5	3.3–$3.2\,S_s$	2.7–$2.0\,S_s$	2.1–$1.2\,S_s$	0.9	0.9

Note: See also Table 11.4-1.

Table GA-2 Interpolation Formulas for Computing Site Class Factor F_v

| Site Class | Value of F_v for a Given Range of Values of S_1 | | | | | |
	(1) <0.10	(2) 0.10–0.20	(3) 0.20–0.30	(4) 0.30–0.40	(5) 0.40–0.50	(6) >0.50
A	0.8	0.8	0.8	0.8	0.8	0.8
B	1.0	1.0	1.0	1.0	1.0	1.0
C	1.7	1.8–S_1	1.8–S_1	1.8–S_1	1.8–S_1	1.3
D	2.4	2.8–$4S_1$	2.4–$2S_1$	2.4–$2S_1$	2–S_1	1.5
E	3.5	3.8–$3S_1$	4–$4S_1$	4–$4S_1$	2.4	2.4

Note: See also Table 11.4-2.

Fig. GA-1

Variation of site factor coefficient F_a with S_s

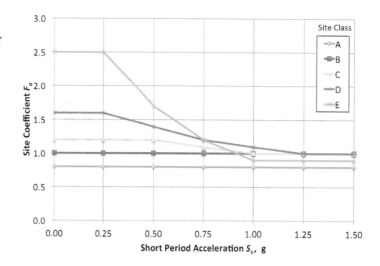

Fig. GA-2

Variation of site factor coefficient F_v with S_1

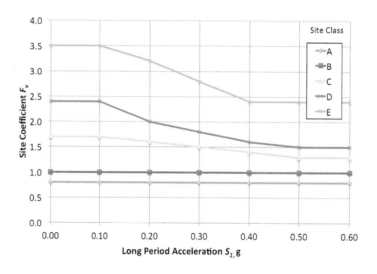

Fig. GA-3

Interpolation functions for C_u (see also Table 12.8-1)

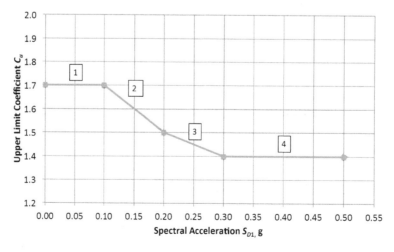

Fig. GA-4

Interpolation functions for exponent *k* (see also Section 12.8.3)

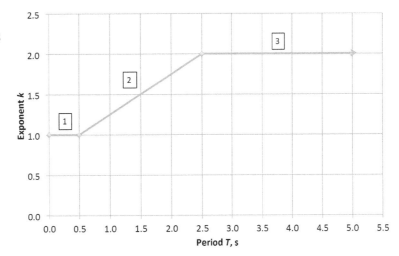

Appendix B

Using the USGS Seismic Hazards Mapping Utility

In ASCE 7-10 the Chapter 22 maps generally do not provide sufficient resolution to determine spectral accelerations, particularly in areas of the country where the contours are closely spaced. For this reason, the legends in the maps (e.g., Figure 22-1) provide an address for a web application that can be used to obtain the spectral accelerations S_s and S_1 and many other design values. Unfortunately, the link provided on the map is not always accurate. At the time of this writing the proper address is http://earthquake.usgs.gov/hazards/designmaps/. If this link does not work a search may be required to find the site.

In the following example the design values for a site near Eugene, Oregon are provided. The coordinates are as follows:

> Latitude = 43.97 (position north-south)
> Longitude = -123.14 (position east-west)

Note that a negative longitude is provided because this site is west of the prime meridian. The structure to be designed is assigned a Risk Category III and is on Site Class D soils.

When the website is loaded, the screen shown in **Fig. GB-1** appears. As seen in the figure, the application provides four drop-down items and a map. When using ASCE 7-10, the design code preference "ASCE 7-10" should be selected from the first drop-down item.

After ASCE 7-10 is specified, the utility changes somewhat, and six drop-down items are provided in addition to the map. This is shown in **Fig. GB-2**.

Fig. GB-1
USGS design value utility when first loaded

Fig. GB-2
USGS design value utility after selecting ASCE 7-10

In this screen, the user enters an optional report title, the site classification, and the risk category. Latitude and longitude can be entered manually, or the mouse can be used to position the cursor within the map and automatically enter the latitude and longitude. A high degree of accuracy may be obtained by using the zoom features in the map. When all values are entered, the utility appears as shown in **Fig. GB-3**.

After all values are provided, the user clicks the "Compute Values" button, and after a few seconds the simple report shown in **Fig. GB-4** is provided. If desired, a much more detailed report can be provided by clicking the "view the detailed report" item at the bottom of the simple report. A portion of the detailed report is shown in **Fig. GB-5**. The detailed report provides all intermediate values used in the determination of the design values and is suitable for inclusion in project calculations.

Fig. GB-3

USGS design value utility after all project data have been entered

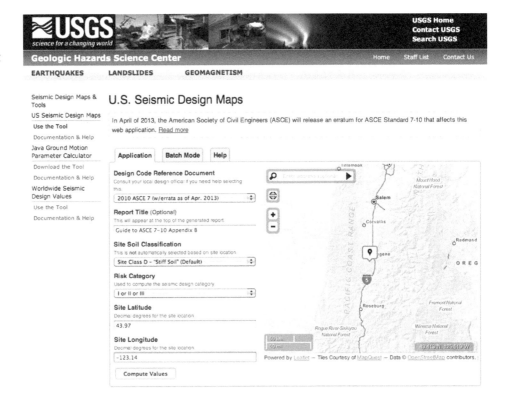

Fig. GB-4
The simple report

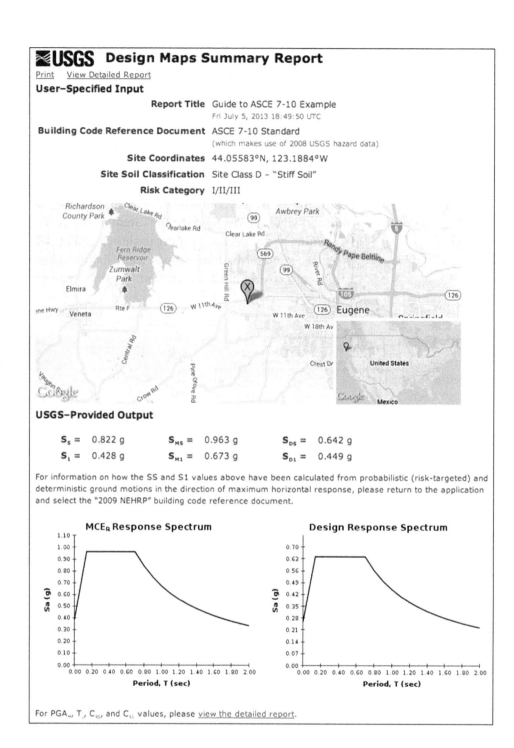

Fig. GB-5

A portion of the detailed report

≋USGS Design Maps Detailed Report

Print View Summary Report

ASCE 7-10 Standard (44.05583°N, 123.1884°W)

Section 11.4.1 — Mapped Acceleration Parameters

Note: Ground motion values provided below are for the direction of maximum horizontal spectral response acceleration. They have been converted from corresponding geometric mean ground motions computed by the USGS by applying factors of 1.1 (to obtain S_s) and 1.3 (to obtain S_1). Maps in the 2010 ASCE-7 Standard are provided for Site Class B. Adjustments for other Site Classes are made, as needed, in Section 11.4.3.

From Figure 22-1	$S_S = 0.822$ g

From Figure 22-2	$S_1 = 0.428$ g

Section 11.4.2 — Site Class

The authority having jurisdiction (not the USGS), site-specific geotechnical data, and/or the default has classified the site as Site Class D, based on the site soil properties in accordance with Chapter 20.

Table 20.3-1 Site Classification

Site Class	$\bar{v}_s$	$\bar{N}$ or $\bar{N}_{ch}$	$\bar{s}_u$
A. Hard Rock	>5,000 ft/s	N/A	N/A
B. Rock	2,500 to 5,000 ft/s	N/A	N/A
C. Very dense soil and soft rock	1,200 to 2,500 ft/s	>50	>2,000 psf
D. Stiff Soil	600 to 1,200 ft/s	15 to 50	1,000 to 2,000 psf
E. Soft clay soil	<600 ft/s	<15	<1,000 psf
	Any profile with more than 10 ft of soil having the characteristics: • Plasticity index $PI > 20$, • Moisture content $w \geq 40\%$, and • Undrained shear strength $\bar{s}_u < 500$ psf		
F. Soils requiring site response analysis in accordance with Section 21.1	See Section 20.3.1		

For SI: 1ft/s = 0.3048 m/s 1lb/ft² = 0.0479 kN/m²

Section 11.4.3 — Site Coefficients and Risk–Targeted Maximum Considered Earthquake (MCE$_R$) Spectral Response Acceleration Parameters

Table 11.4-1: Site Coefficient F_a

Site Class	Mapped MCE$_R$ Spectral Response Acceleration Parameter at Short Period				
	$S_S \leq 0.25$	$S_S = 0.50$	$S_S = 0.75$	$S_S = 1.00$	$S_S \geq 1.25$
A	0.8	0.8	0.8	0.8	0.8
B	1.0	1.0	1.0	1.0	1.0
C	1.2	1.2	1.1	1.0	1.0
D	1.6	1.4	1.2	1.1	1.0
E	2.5	1.7	1.2	0.9	0.9
F	See Section 11.4.7 of ASCE 7				

Note: Use straight–line interpolation for intermediate values of S_S

Appendix C

Using the PEER NGA Ground Motion Database

This document explains how to use the PEER Ground Motion Database. In most cases, this database is used to download ground motion record sets to be used in response history analysis of structures.

To access the database, go to the main page for the site, http://peer.berkeley.edu/nga/.

To get ground motion records, click on "Search" (in the middle of the second row from the top). The website will take you to a form with a map on the left side of the screen and various dropdown lists on the right.

To find a specific earthquake, make your selection from the dropdown box next to "Earthquake." For example, to find the Loma Prieta earthquake, scroll down to "Loma Prieta 1989-10-18 00:05" and select it.

To see where the records are located, click the "Search" button at the bottom of the form. The results will appear in a map. After zooming in on the map, you will see something like the map shown in **Fig. GC-1**. If you click on "Monterey," for example, you will see the map in **Fig. GC-2**.

Alternatively, rather than use the map, you may obtain a list of all ground motion records related to the earthquake. To do this, change "Display Results" from "on Map" to "in Table." You will see a long list of record sets, only part of which is shown below:

Record	Earthquake	Station
NGA0731	Loma Prieta 1989-10-18 00:05 (6.93)	CDMG 58373 APEEL 10—Skyline
NGA0732	Loma Prieta 1989-10-18 00:05 (6.93)	USGS 1002 APEEL 2—Redwood City
NGA0733	Loma Prieta 1989-10-18 00:05 (6.93)	CDMG 58393 APEEL 2E Hayward Muir
NGA0734	Loma Prieta 1989-10-18 00:05 (6.93)	CDMG 58219 APEEL 3E Hayward CSUH
NGA0735	Loma Prieta 1989-10-18 00:05 (6.93)	CDMG 58378 APEEL 7—Pulgas
NGA0736	Loma Prieta 1989-10-18 00:05 (6.93)	USGS 1161 APEEL 9—Crystal Springs Res
NGA0737	Loma Prieta 1989-10-18 00:05 (6.93)	CDMG 57066 Agnews State Hospital

Fig. GC-1
Map for Loma Prieta earthquake

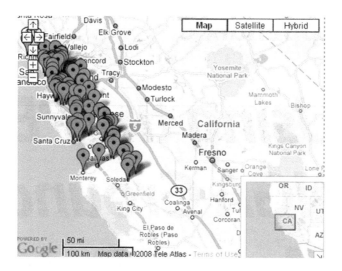

Fig. GC-2
Selecting Monterey City Hall record

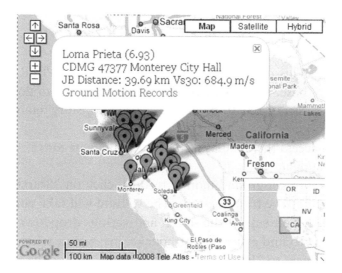

Return to the map-based search display. Using the example search for Monterey, click on "Ground Motion Records" in the bubble caption, and the screen shown in **Fig. GC-3** appears.

This diagram gives information about the earthquake and links to get the actual ground motion acceleration records. There are usually three records per set: north–south (in this case 000), east–west (090), and vertical (UP) records.

To see the north–south ground motion record, click on the record name, for example, LOMAP/MCH000 (displayed in the map-based search results). This is the north–south component because the compass bearing is given as "000" in the title. Compass bearings are shown in **Fig. GC-4**.

The following data will then appear (with many more lines of data than shown below):

Fig. GC-3

Detailed information about single ground motion

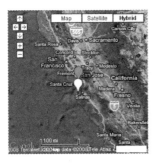

Earthquake: Loma Prieta 1989-10-18 00:05
Magnitude: 6.93
Mo: 2.7861E+26
Mechanism: 3
Hypocenter Latitude: 37.0407 | Longitude: -121.883 | Depth: 17.5 (km)
Fault Rupture Length: 40.0 (km) | Width: 18.0 (km)
Average Fault Displacement: 108.1 (cm)
Fault Name: San Andreas-Santa Cruz
Slip Rate: 17.00 (mm/yr)

Station: CDMG 47377 Monterey City Hall
Latitude: 36.5970 | Longitude: -121.897
Geomatrix 1: A | Geomatrix 2: G | Geomatrix 3: A
Preferred Vs30: 684.90 (m/s) | Alt Vs30:
Instrument location: BASEMENT

Epicentral Distance: 49.39 (km) | Hypocentral Distance: 52.39 (km) | Joyner-Boore Distance: 39.69 (km)
Campbell R Distance: 44.35 (km) | RMS Distance: 52.51 (km) | Closest Distance: 44.35 (km)
PGA: 0.0700 (g)
PGV: 4.5400 (cm/sec)
PGD: 2.0900 (cm)

ATH	PGA (g)	PGV (cm/s)	PGD (cm)	Filter	nPass	nRoll	HP	LP	Lowest Usable Frequency
LOMAP/MCH000				C	1		0.2	28	0.25
LOMAP/MCH090							0.2	22	0.25
LOMAP/MCH-UP									

Fig. GC-4

Compass for ground motion bearings

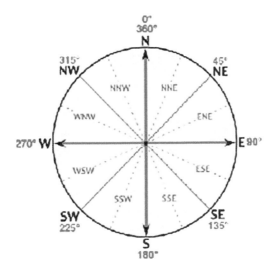

```
PEER NGA STRONG MOTION DATABASE RECORD
LOMA PRIETA 10/18/89 00:05, MONTEREY CITY HALL, 000 (CDMG STATION 47377)
ACCELERATION TIME HISTORY IN UNITS OF G
7990      0.0050      NPTS, DT
  -0.129203E-02  -0.155257E-02  -0.117803E-02  -0.355531E-03  -0.482559E-03
  -0.484030E-03  -0.481456E-03  -0.477630E-03  -0.476219E-03  -0.479441E-03
  -0.487404E-03  -0.499153E-03  -0.513798E-03  -0.531229E-03  -0.551795E-03
  -0.574890E-03  -0.598821E-03  -0.621665E-03  -0.641146E-03  -0.654379E-03
  -0.658443E-03  -0.651863E-03  -0.635102E-03  -0.610199E-03  -0.580759E-03
  -0.552201E-03  -0.529093E-03  -0.511311E-03  -0.493790E-03  -0.469597E-03
  -0.434634E-03  -0.388477E-03  -0.334624E-03  -0.280486E-03  -0.230926E-03
  -0.187375E-03  -0.152341E-03  -0.129905E-03  -0.122443E-03  -0.129574E-03
  -0.150408E-03  -0.182927E-03  -0.222825E-03  -0.261722E-03  -0.288505E-03
  -0.298566E-03  -0.297028E-03  -0.290274E-03  -0.284506E-03  -0.281662E-03
  -0.276887E-03  -0.266221E-03  -0.251237E-03  -0.242671E-03  -0.249410E-03
  -0.272418E-03  -0.313515E-03  -0.368816E-03  -0.425876E-03  -0.474587E-03
```

The ground motion data is written in rows. This file contains 7,990 data points, written at a time increment of 0.005 s. The accelerations are written in *g* units.

To save the data to your computer, "Select All" of the data, copy, and then paste into a NOTEPAD file.

To use the data in analysis software, you may need to copy the four lines of header information that accompanies each record.

Other information: The PEER NGA database also contains a huge Excel spreadsheet called the "flatfile," which contains a host of information about each ground motion. This spreadsheet can be downloaded from the "Download" link at the top of the file, and the explanation of the data in the spreadsheet can be obtained from the "Documentation" link. This information will be useful when selecting several earthquakes with similar characteristics, such as fault type, magnitude, site characteristics, and so on.

References

American Concrete Institute (ACI). (2008). "Building code requirements for structural concrete," *ACI 318-08* and *ACI 318-08R*, Farmington Hills, MI.

American Institute of Steel Construction (AISC). (2005). "Seismic provisions for structural steel buildings," *AISC/ANSI 341-5*, Chicago.

American Institute of Steel Construction (AISC). (2010a). "Specification for structural steel buildings," *AISC/ANSI 360-10*, Chicago.

American Institute of Steel Construction (AISC). (2010b). "Seismic provisions for structural steel buildings," *AISC/ANSI 341-10*, Chicago.

Applied Technology Council (ATC). (1984). "Tentative provisions for the development of seismic regulations for new buildings," *Report ATC 3-06*, Redwood City, CA.

ASCE. (2007). "Seismic rehabilitation of existing buildings," *ASCE 41-06*, Reston, VA.

ASCE. (2013). "Minimum design loads for buildings and other structures," *ASCE/SEI 7-10*, Reston, VA. [Note: this refers to the third printing, including supplement.]

ASTM. (2005a). "Standard test methods for laboratory determination of water (moisture) content of soil and rock by mass," *ASTM D2216-05*, West Conshohocken, PA.

ASTM. (2005b). "Standard test methods for liquid limit, plastic limit, and plasticity index of soils," *ASTM D4318-05*, West Conshohocken, PA.

ASTM. (2006). "Standard test method for unconfined compressive strength of cohesive soil," *ASTM D2166*, West Conshohocken, PA.

ASTM. (2007). "Standard test method for unconsolidated-undrained tri-axial compression test on cohesive soils," *ASTM D2850-03a*, West Conshohocken, PA.

ASTM. (2008). "Standard test method for standard penetration test (SPT) and split-barrel testing," *ASTM D1586*, West Conshohocken, PA.

ASTM. (2010). "Standard practice for classification of soils for engineering purposes (Uniform Classification System)," *ASTM D2487*, West Conshohocken, PA.

Aswegan, K., and Charney, F. A. (2014). "A simple linear response history analysis procedure for building codes," *Proceedings of the 10th U.S. Conference on Earthquake Engineering*, Anchorage, AK.

Charney, F. A. (1990). "Sources of elastic deformations in tall steel frame and tube structures," *Proceedings of the Conference Tall Buildings, 2000 and Beyond*, Tall Building Council, Hong Kong.

Charney, F. A., and Marshall, J. (2006). "A comparison of the Krawinkler and scissors models for including beam–column joint deformations on the analysis of moment-resisting frames." *Eng. J. (AISC)*, 43(1), 31–48.

Charney, F.A., Talwalker, R., Bowland, A., and Barngrover, B. (2010). "NONLIN-EQT, a computer program for earthquake engineering education," *Proceedings of the 10th Canadian and 9th U.S. Conference on Earthquake Engineering*, Toronto, ON, Canada.

Chopra, A. K. (2011). *Dynamics of structures*, 4th ed., Prentice-Hall, Upper Saddle River, NJ.

CSI. (2009). *SAP 2000 theoretical manual*, Computers and Structures, Berkeley, CA.

FEMA. (2009a). "NEHRP recommended seismic provisions for new buildings and other structures," *FEMA P-750*, Washington, D.C.

FEMA. (2009b). "Quantification of building seismic performance factors," *FEMA P-695*, Washington, D.C.

FEMA. (2010). "Earthquake resistant design concepts," *FEMA P-749*, Washington, D.C.

FEMA. (2012). "NEHRP recommended provisions: Design examples," *FEMA P-751*, Washington, D.C.

Ghosh, S. K., and Dowty, S. (2007). "Bearing wall systems vs. building frame systems," *Go Structural*, <http://www.gostructural.com/article.asp?id=1558> (accessed July 16, 2009).

Goel, R.K., and Chopra, A.K. (1997). "Period formulas for concrete shear wall buildings," *J. Struct. Eng.*, 123(11), 1454-1461.

Hancock, J., Watson-Lamprey, J., Abrahamson, N., Bommer, J., Markatis, A., McCoy, E., and Mendis, R. (2006). "An improved method of matching response spectra of recorded earthquake ground motion using wavelets." *Journal of Earthquake Engineering*, 10(Special Issue 1), 67–89.

International Code Council (ICC). (2011). *2012 International building code*, Country Club Hills, IL.

Liew, J. Y. (2001). "Inelastic analysis of steel frames with composite beams." *J. Struct. Eng.*, 127(2), 194–202.

National Institute of Standards and Technology (NIST). (2010a). "Nonlinear structural analysis for seismic design," *Report NIST GCR 10-917-6*, Gaithersburg, MD.

National Institute of Standards and Technology (NIST). (2010b). "Evaluation of the FEMA P-695 methodology for quantification of building seismic performance factors," *Report NIST GCR 10-917-8*, Gaithersburg, MD.

National Institute of Standards and Technology (NIST). (2011). "Selecting and scaling earthquake ground motions for response history analysis," *Report NIST GCR 11-917-15*, Gaithersburg, MD.

Park, R., and Paulay, T. (1974). *Reinforced concrete structures*, Wiley Interscience, New York.

Paulay, T., and Priestly, M. J. N. (1992). *Seismic design of reinforced concrete and masonry structures*, Wiley, New York.

PTC. (2012). Mathcad Engineering Calculations Software, PTC Inc., Needham, MA.

Rack Manufacturers Institute (RMI). (2009). *Specifications for the design, testing, and utilization of industrial steel storage racks*, Charlotte, NC.

Sabelli, R., Pottebaum, W., and Dean, B. (2009). "Diaphragms for seismic loading." *Structural Engineer*, January.

Schaffhausen, R., and Wegmuller, A. (1977). "Multistory rigid frames with composite girders under gravity and lateral forces." *Eng. J. (AISC)*, 2nd Quarter.

Wilson, E. L. (2004). *Static and dynamic analysis of structures*, Computers and Structures, Berkeley, CA.

Index

characteristic site period, computation of, 80–81, 81*e*, 81*t*

column systems; horizontal irregularities and overstrength factor requirement, 190–91; in-plane discontinuity, overturning moment demands, and overstrength factor requirement, 68–69, 69*f*; story stiffness analysis and beam-to-column stiffness ratios, 186–88, 186*f*, 188*f*; story strength computation, 189–190, 189*f*, 190*e*; vertical irregularities and overstrength factor requirement, 190–91

combination systems; approximate period of vibration, 194; story strength computations, 70–71; vertical combinations of structural systems, 48–50, 48*f*; vertical combinations when lower section is stiffer than upper section, 51

combined scale factor, 32*f*, 33–34

complete quadratic combination (CQC); modal combination with, 163; MRS analysis and diaphragm forces, 197; MRS analysis and load combination procedures, 102–3; signs of member forces, 198–99

composite slabs, 120

concentrically braced frame systems; dual systems with concentrically braced frame, 39*t*, 40*t*; special steel concentrically braced frame structural system, 48–51, 48*f*; specifications for special concentrically braced frame structural system, 39*t*, 40*t*; story strength computation, 70–71, 71*f*; structural analysis procedures for, 81, 81*t*; two-stage ELF procedure, 132–34, 132*f*, 133*f*

concrete shear wall systems; bearing wall systems, 41–43, 42*f*, 43*f*; computing approximate period for, 120–22, 120*e*, 121*f*, 121*t*; cracking and shear cracking, 91, 193; design parameters, 128; ELF analysis of five-story building, 127–132, 129*f*, 129*t*, 130*e*, 131*t*; shear deformations, 120; special reinforced concrete shear walls, 45–48, 46*f*; torsional loading evaluation, 89–96, 90*f*, 91*t*, 92*f*, 93*t*, 94*t*, 95*f*, 95*t*; two-stage ELF procedure, 132–34, 132*f*, 133*f*; vertical combinations of structural systems, 51

concrete structures; concrete flat slab structural system and effective seismic weight of four-story warehouse, computation of, 105–12, 106*f*, 107*f*, 112*t*; concrete slab diaphragm forces example, 181–84, 182*f*, 183*t*; cracking and computing cracked section properties, 82, 193. *See also* reinforced concrete systems

courthouse and office building Risk Category exercise, 3

custodial care facilities; elder-care facilities Risk Category exercise, 3; risk category and occupancy group, 1

day care facility risk category and occupancy group, 2

dead load; computation of, 108–11; design dead load, 107; drift computation, 135, 136*t*; weight computation and, 108, 112*t*

deflection, computation of, 54*e*

deflection amplification factor, 37–38, 196–97

design response spectrum, 25–27, 26*f*, 27*t*

detention center risk category and occupancy group, 2

diaphragms; analytical procedure to determine flexibility, 73–77, 74*f*, 75*f*, 76*f*; average drift of vertical element (ADVE), 74–75, 75*f*, 76–77, 76*f*; chord elements and forces, 182, 182*f*, 183, 197; collector elements and forces, 182, 182*f*, 183, 197; discontinuity irregularity, 57–60, 58*f*, 59*f*; flexible, 73, 74; flexible semirigid, 95–96, 95*f*; inertial forces, 182–83, 182*e*; interconnection requirement, 185; maximum diaphragm deflection (MDD), 74–75, 75*f*, 76–77, 76*f*; modeling requirements and methods, 81–82; MRS analysis and diaphragm forces, 184, 197; nodal forces, 95–96, 95*f*; in-plane deformations, 90–91; in-plane forces, 181–84, 182*f*, 183*t*; rigid, 73, 82; section cuts and interconnection requirement, 185; semirigid, 73, 77–78, 82, 90; semirigid ELF analysis, 89–96, 90*f*, 91*t*, 92*f*, 93*t*, 94*t*, 95*f*, 95*t*; span of, determination of, 73; span-to-depth ratio, 73; stiffness calculations, 59, 59*f*; stiffness of, 59–60

Northridge earthquake ground motion
record set, 30–36, 30*t*, 31*f*, 32*f*, 35*f*,
168–170, 168*t*, 170*f*, 171*f*

occupant loads, 2
office buildings, 2, 4
orthogonal load; eccentricity and
application of, 102; ELF load case
generation, 101–2; inclusion of in load
requirements, 131; MRS analysis and
load combination procedures, 102–3
out-of-plane offset irregularity, 60, 60*f*
overstrength; computation of, 192;
computation of actual story
overstrengths, 144–45, 144*e*; factors
that contribute to, 141; requirements
for, 140*t*, 141, 145*t*
overstrength factor, 37–38, 196–97;
application of, 197–98; characteristics
of elements designed using, 197–98;
collector element design, 183; load
combinations and, 98, 104; in-plane
discontinuity, overturning moment
demands, and, 50, 68–69, 69*f*
overturning forces, reduction of,
131–32
overturning moments; amplitude-scaled
ground motions, 175, 175*t*; MRH
analysis, 173, 175, 175*t*, 177, 178*t*;
in-plane discontinuity, overstrength
factor and, 68–69, 69*f*; spectrum-
matched ground motions, 177, 178*t*

Pacific Earthquake Engineering Research
Center (PEER); flatfile
spreadsheet, 214; Ground Motion
Database instructions, 211–14, 212*f*,
213*f*; next-generation attenuation
(NGA) record set, 30–31; PEER
Strong Motion Database, 30
P-delta effects; basis for calculation
of, 191–92; computer analysis for
structural analysis and, 192–93; drift
computation, 137, 140–43, 141*t*, 142*t*,
192–93; equations that include, 137;
hazard risk and limits on, 192;
inclusion requirement, 82; live loads
and P-delta analysis, 135; stability
computation and, 191–93; stability
ratio calculations, 140–45, 140*e*, 143*t*,
145*t*
PEER. *See* Pacific Earthquake Engineering
Research Center (PEER)
period of vibration, 115–122; approximate
period of vibration, 50–51, 115–122,
115*e*, 116*f*, 118*e*, 118*t*, 119*t*, 194;

computed period, 117–122, 118*e*,
118*t*, 119*e*, 119*t*; computer programs
for period computed, 119–120;
displacement computation, 154–55,
155*t*; drift computation, 138–140,
138*e*, 138*t*, 139*t*, 154–55, 155*t*, 192;
equation for estimating, 90*f*; exponent
k interpolation, 203*f*; masonry and
concrete shear wall structures,
computing approximate period
for, 120–22, 120*e*, 121*f*, 121*t*;
three-dimensional systems,
determination of period for, 122;
torsional amplification and, 90;
torsional irregularity evaluation
and, 90
piecewise exact method, 172
prisoner holding cells Risk Category
exercise, 3

radio dispatch facilities Risk Category
exercise, 3
redundancy factor; calculation test, 84;
configuration test, 84; determination
of, 83–87, 84*f*, 86*f*, 87*f*, 98; elastic
analysis, 85–87, 86*f*; inelastic
analysis, 85–87, 87*f*; requirement
for, 191
reentrant corner irregularity, 57, 58, 58*f*,
59*f*, 90, 90*f*
reinforced concrete systems; cracking
and computing cracked section
properties, 120, 193; design
parameters, 128; ELF analysis of
five-story building, 127–132, 129*f*,
129*t*, 130*e*, 131*t*; period computed
using computer programs, 120; special
reinforced concrete moment
frames, 45–48, 46*f*, 195
response history analysis; ground motion
scaling procedure, 30*t*, 31*f*, 32*f*, 35*f*.
See also linear response history (LRH)
analysis; modal response history
(MRH) analysis; nonlinear response
history (NRH) analysis
response modification coefficient, 37–41,
196–97
retail facilities (shops and restaurants) Risk
Category exercise, 3–4
Risk Category, 1–6; descriptions of
categories, 1–2; exercises, 2–6;
importance factor and, 7–8, 8*f*;
Occupancy Groups, 1–2; occupant
loads, 2; Seismic Design Category
and, 8–10, 9*t*
RSP-Match computer program, 169

storage facilities; computation of effective seismic weight of four-story warehouse, 105–12, 106f, 107f, 112t; dockside cargo storage warehouses Risk Category exercise, 5; grain storage silos Risk Category exercise, 6; hazardous chemical storage facility, 73–77, 74f; live load and weight computation, 197

story displacements; computation of, 54, 54e, 127, 127f, 154–55, 155t, 192; MRH analysis, 173; period of vibration and computation of, 154–55, 155t; ratio at edge to center, 54e; ratio of edge to end, 55f; torsional irregularity evaluation, 54–57, 56f, 56t, 93, 93t

story drifts; amplitude-scaled ground motions, 174, 174t, 178t; computation of, 135–145, 136f, 136t, 138t, 139t, 142t, 143t, 154–55, 155t, 192, 194–95; dead loads and computation of, 135, 136t; definition of, 194; ductility demand and, 8; ELF, MRS, and MRH results comparison, 178t; lateral forces, computed period, and computation of, 127; lateral loads and computation of, 135–36; live loads and computation of, 135, 136t; MRH analysis, 174, 174t, 176–77, 176t; P-delta effects and computation of, 137, 140–43, 141t, 142t, 192–93; period of vibration and computation of, 138–140, 138e, 138t, 139t, 154–55, 155t, 192; period values for use in calculations, 118t; ratio of edge to center, 53, 56t, 57; shear deformations, 120, 194; soft story irregularity drift-based check, 63–66, 66t, 67, 67f; spectrum-matched ground motions, 176–77, 176t, 178t; structural system characteristics and, 40; torsional irregularity evaluation, 53–57, 56f, 56t, 92–93

story forces, 155–56, 155e, 156t, 172–73, 172e

story shears, 155–57, 156t; amplitude-scaled ground motions, 174–75, 175t, 179t; ELF, MRS, and MRH results comparison, 179t; MRH analysis, 173, 174–75, 175t, 177, 177t; spectrum-matched ground motions, 177, 177t, 179t

story stiffness; analysis of story stiffness and beam-to-column stiffness ratios, 186–88, 186f, 188f; hazard risk

and P-delta effects, 192; period computed using computer programs, 119–120; soft story irregularities and stiffness-based check, 63, 66–67, 67t; stability computation and P-delta effects, 191–93; two-stage ELF procedure, 132–34, 132f, 133f; vertical structural irregularities, 186–88, 186f, 188f

story strength; beam mechanism, 70, 70e, 70f, 189f, 190, 190e; braced frame systems, 70–71, 71f, 189, 189e, 189f; buckling-restrained braced frame structural system, 70–71, 71f; column mechanism, 189–190, 189f, 190e; computation of, 144, 145, 188–190, 189f; concentrically braced frame systems, 70–71, 71f; frame-wall systems, 70–71; moment frame systems, 69–70, 70e, 70f, 189–190, 189f, 190e; period values for use in calculation of, 118t; plastic story strength, 144; strong column–weak beam design rule, 141, 144

structural analysis; diaphragm forces analysis, 183; modeling requirements, 81–82, 148; selection of procedure, 79–82

structural integrity, requirements to meet General Structural Integrity provisions, 185

structural systems; drift and, 40; factors in selection of, 37; four-story and ten-story buildings example, 37–41, 38f; members or connections, design of and load combinations, 97–104, 99f, 100f, 102t, 103f; selection of and restrictions on types and heights for seismic resistant designs, 37–41, 39t, 40t; vertical combinations of, 48–50, 48f

symbols, ix, ix–x

Tennessee site Seismic Design Category, 8–10, 9t

three-dimensional modal response spectrum (MRS) analysis, 161–64, 161f, 162t, 163f, 164t

torsional amplification factor, computation of, 55, 55e, 56t, 57, 57e, 93, 93e, 94–95, 94t, 95t

torsional irregularities; building design and probability of, 131; definition of, 53; evaluation of, 53–57, 54f, 55f, 56f, 56t; extreme torsional irregularity, 53, 54; load combinations and effects

About the Author

Finley A. Charney, Ph.D., P.E., F.ASCE, F.SEI is professor of civil and environmental engineering in the Charles E. Via, Jr., Department of Civil and Environmental Engineering at Virginia Tech. He also maintains a professional practice through his consulting firm, Advanced Structural Concepts, Inc., located in Blacksburg, Virginia. He has 21 years of experience as a practicing engineer and 13 years of academic experience. He received his B.S. in Civil Engineering and his M.S. in Architectural Engineering from the University of Texas at Austin. He earned his Ph.D. in Engineering from the University of California at Berkeley. His area of research is structural analysis, structural dynamics, earthquake engineering, and structural system development.

Dr. Charney is active in the area of earthquake engineering education and has developed and presented numerous professional seminars in the field. These include the ASCE Continuing Education courses, "Fundamentals of Earthquake Engineering" and "Seismic Loads for Buildings and Other Structures." He is a member of the ASCE 7 Seismic Loads Committee and served on task committees for updating the *NEHRP Recommended Provisions for Seismic Regulations for New Buildings and Other Structures*. He is a registered professional engineer in the states of Texas and Colorado.